TOWER HAMLETS

KT-473-496

91 000 004 140 54 7

Library Learning Information

To renew this item call:

0115 929 3388

or visit

www.ideastore.co.uk

TOWER HAMLETS

Created and managed by Tower Hamlets Council

THE END

OF

PLAGUES

THE END
OF
PLAGUES

THE GLOBAL BATTLE AGAINST
INFECTIOUS DISEASE

JOHN RHODES

palgrave
macmillan

THE END OF PLAGUES
Copyright © John Rhodes, 2013.
All rights reserved.

First published in 2013 by PALGRAVE MACMILLAN® in the U.S.—a
division of St. Martin's Press LLC, 175 Fifth Avenue, New York, NY
10010.

Where this book is distributed in the UK, Europe and the rest of the
world, this is by Palgrave Macmillan, a division of Macmillan Publishers
Limited, registered in England, company number 785998, of Houndmills,
Basingstoke, Hampshire RG21 6XS.

Palgrave Macmillan is the global academic imprint of the above
companies and has companies and representatives throughout the world.

Palgrave® and Macmillan® are registered trademarks in the United
States, the United Kingdom, Europe and other countries.

ISBN: 978-1-137-27852-4

The Library of Congress has catalogued the hardcover edition as follows:

Rhodes, John, 1947–
  The end of plagues : the global battle against infectious disease / John
Rhodes.
    p.   ;   cm.
  ISBN 978-1-137-27852-4
  I. Title.
  [DNLM: 1. Virus Diseases—history. 2. Virus Diseases—prevention &
control. 3. Allergy and Immunology—history. 4. Disease Eradication—
history. 5. Vaccination—history. 6. World Health. WC 11.1]
  RC114.5
  362.1969'1—dc23

                                                              2013007383

A catalogue record of the book is available from the British Library.

Design by Letra Libre, Inc.

First edition: September 2013

10 9 8 7 6 5 4 3 2 1

Printed in the United States of America.

To my wife, Heather

TOWER HAMLETS LIBRARIES	
91000004140547	
Bertrams	31/10/2013
362.196	£17.99
THISBO	TH13000827

CONTENTS

Foreword by Sir Richard Sykes   ix

Introduction   1

1   The Power of the Invisible   5

2   Circassian Beauties and Pioneering Women   11

3   The Making of Jenner   23

4   Why Not Try the Experiment?   29

5   The Fourth Achievement   37

6   The Foundling Voyages   47

7   The Teeming Humanity of Nations   55

8   A Great and Loud Commotion   65

9   Completing the Picture   71

10 Germ Theory and the Birth of Immunology   77

11 Victorious Weapons against Illness and Death   89

12 First Light on the Mystery of Infantile Paralysis   101

13 Yearning to Breathe Free   109

14 A Great Step Forward   123

15 Great Themes and Dirty Little Secrets   139

16 The War on Influenza   143

17 Forged in the Crucible of War   151

18 Smallpox in a Land of Ancient Wisdom   159

19 The Final Defeat of Smallpox   167

20 Invisible Weapons of War   173

21 Benefits, Risks, and Fears   179

22 Inspiration in the Global Village   191

23 A Team of Many Colors   199

24 The Milkmaid and the Cuckoo   215

   Acknowledgments   219
   Bibliography   221
   Index   227

Eight pages of photographs appear between pages 108 and 109.

FOREWORD

Considering the nearly infinite variety of microorganisms that have developed over the past 3.5 billion years, relatively few microbes cause disease in otherwise healthy individuals. However, a number of them have evolved into opportunistic pathogens that were the scourges of antiquity—plague, tuberculosis, leprosy, cholera, and syphilis. For most of recorded history the average life expectancy has been between 20 and 40 years, a direct result of the ravages of infectious disease. With the exception of the black plague and tuberculosis, most of the great plagues are caused by viruses—smallpox, polio, influenza, and AIDS.

Unlike bacteria, viruses are incapable of self-replication and need to live and multiply within living cells. First identified in the 1890s, about 5,000 viruses have been discovered to date and, like bacteria, occupy all parts of the ecosystem. Like their bacterial counterparts, the viral plagues have also played a significant role in our history. Following the Great War, the flu pandemic of 1917–18 killed an estimated 50 million people.

Unlike the story of bacteria and the discovery of penicillin, there has been no big eureka moment with viruses. It has been a 250 year journey of trial and error, powerful personalities, great scientific discoveries, and many deaths. This is the story told by John Rhodes.

The earliest observations that the body could be induced to resist infection were made by Thucydides, who, during the great plague of Athens in 430 BC, noted that people who had recovered from a previous bout of the disease could nurse the sick without contracting the illness a second time. By the mid-eighteenth century the plague has disappeared from Europe, only to be replaced with smallpox. In 1770 when Jenner began his work on a vaccine to protect against smallpox, he was dealing with an unknown pathogen and a misunderstood defence mechanism. He relied solely on observations that the body could be induced to resist infections. The nature of the smallpox virus was to remain a mystery until 1906, and the role of the immune system would not be fully understood until well into the twentieth century.

Vaccination is the process by which an individual's immune system becomes fortified against an invading organism, the antibodies resulting from a vaccine fighting against the infection or preventing it altogether. Before vaccines, the only way people became immune to a certain disease was by contracting the disease and hopefully surviving it. Vaccines are composed of either live attenuated or killed bacteria or viruses in order to mimic the virulent form of the disease and trick the body into mounting an immune response.

Although vaccination against smallpox quickly established itself and spread across the world, the nineteenth century was a battleground between theorists over the nature of infectious diseases and the body's response to it, in other words immunity. It is important to remember that this was a period when nothing was known about the microbial basis of infectious diseases and the concept of spontaneous generation was still regarded as the route through which new life forms appeared. Maladjustment of the four humours was blamed for most illnesses, despite the long standing appreciation that some diseases were contagious and that limited exposure could protect against subsequent illness.

In the late nineteenth century, with the identification of microorganisms and a greater understanding of their relationship to disease, the beginnings of microbiology and fundamental immunology were ushered in by Pasteur and Koch. Many bacterial diseases were defined for the first time and the organisms responsible identified. Advances in the study of the epidemiology of infection and host-pathogen relationships led to improved public health policies, infection control, and hygiene and sanitation initiatives. However, a sense of helplessness remained until the introduction of penicillin in the early 1940s. In the period immediately following a large number of antibiotics were discovered, leading many to believe that infections caused by bacteria would soon be a thing of the past. However, bacteria reproduce every 20–30 minutes and have an amazing ability to adapt to environmental changes. Today there are no classes of antibiotics that bacteria have not developed a resistance to. Viruses have also continued to evolve resistance to anti-viral agents.

I trained as a microbiologist and spent a significant part of my life working on developing new antibiotics and studying how bacteria develop a resistance to them. From research and development I moved into management and eventually became the CEO and Chairman of the pharmaceutical company GlaxoWellcome. This is where I first met John

Rhodes. As the holder of the Rock Carling Fellowship in 2000, I wrote a monograph on new medicines, the practice of medicine and public policy. John provided extremely helpful input for the sections on vaccines, and it was this work that inspired him to write this book, *The End of Plagues*.

The story of the great plagues has been told many times, but in this book John brings the story to life by describing the characters, their trials and tribulations, and the issues, many of which continue to haunt us to this day. The prejudice against the whole concept of vaccination has caused and continues to cause enormous suffering. Statements made by people in positions of responsibility advising against vaccination or denying that a disease is caused by infective agents have led to unnecessary pain and death. Vaccines are prophylactics, which protect against disease and are therefore given to healthy people. Antibiotics are usually given to people with an active disease and as a result the risk to benefit ratio is seen very differently. It is essential that people understand that the enormous benefits of vaccination far outweigh the risks of side effects.

As I write this piece, 900 people in South Wales have been diagnosed with measles, a viral disease that can cause blindness and death. This serious outbreak resulted from parents' resistance to immunizing their children with the MMR vaccine due to misinformation and a lack of understanding of the risks versus the benefits of vaccination.

But Jenner's work was not in vain, two hundred years following his pioneering work, smallpox was eliminated from the earth—a major triumph for the human race. Although a great triumph, we cannot be complacent; infectious diseases are still alive and strong and a threat to mankind. The next target for elimination is polio, which poses a much greater challenge than smallpox, but the determination and the commitment are there.

So, what of the future? In today's hospitals we struggle with many drug-resistant pathogens. Bombarded by inimical chemicals for the past 70 years, bacteria have evolved many ingenious ways to develop resistance to them. In recent years the development of new antibiotics has slowed to a trickle—the outlook is not bright. We have had to revert back to the surveillance and containment strategies used in the nineteenth century along with the careful use of precious antibiotics.

Producing vaccines against viral threats is a potentially hazardous business and that's why modern manufacturers have to operate strict controls to ensure that no pathogens escape if live vaccines are being used, as is the case with the polio vaccine. One way to avoid this problem is to move away from live virus vaccines to what are called subunit vaccines. A great deal of research is now being carried out to develop new methods of creating entirely synthetic vaccines that do not rely on using live infectious virus.

We need to better understand the characteristics of infection. Epidemics require organisms to be infective and infectious. Infectivity describes the ability of a particular pathogen to enter, survive, and multiply in the host, while the infectiousness of a disease is the comparative ease with which disease is transmitted to others. The human strains of Ebola virus, for example, incapacitate its victims extremely quickly and kills them soon after. As a result, the victims of this disease do not have the opportunity to travel very far from the initial infection zone. The spread of Ebola is very rapid and usually stays within a relatively confined geographical area. In contrast, the human immunodeficiency virus (HIV) kills its victims slowly by attacking their immune system. As a result, many of its victims transmit the virus to other individuals before even realizing that they are carrying the disease. Also, the low virulence allows its victims to travel long distances, increasing the likelihood of an epidemic.

Over 3.5 billion years of evolution has developed one of the most complex systems in biology, the immune system, without which we would not exist. In *The War of the Worlds* by H. G. Wells the invaders from Mars were not destroyed by the instruments of war, but instead by simple microorganisms. Why? Because they had no immune system.

—*Sir Richard Sykes*

INTRODUCTION

Evolutionary biologists tell us that life began on earth around 4 billion years ago but that humans arrived late on the scene. If the history of life on our planet were to span a single day, our hominid relatives would not appear until the last minute. At the beginning of that day, simple but successful creatures ruled, and some of these evolved into infectious pathogens that cause disease. Modern humans have been around for 200,000 years, but only in the last two centuries have we come to learn the truth: if our species' time on earth was represented by a single day, we'd know about germs only in the last two minutes. And for all that time we were ignorant, as well, about our natural defenses against them. It is only in the last minute of our single day on earth that the mysteries of immune defense have been dispelled.

This book is about the universe of germs and the nature of immunity. It explores the natural ways that we defend ourselves against disease, as well as the increasingly sophisticated ways the immune system can be harnessed to combat the most deadly infections. In our twenty-first-century world, we already possess more than 25 vaccines against serious diseases, smallpox has been eradicated, and we can now contemplate the imminent eradication of polio. We can even begin the global fight against killers like malaria, tuberculosis, and HIV/AIDS.

The End of Plagues begins with Edward Jenner's discovery of vaccination, the first chapter in a modern story after centuries of futile classical notions about disease and its causes. But Jenner's discovery did not jump out of nowhere: in eighteenth-century England an ingenious and sometimes dangerous technique to prevent smallpox had already been established. At a time when medical measures for "the Small Pox"—"the most terrible of all the ministers of death"—were confined to vomiting, purging, blistering, and bloodletting, two pioneering, highborn women championed variolation, the hazardous forerunner to Jenner's discovery. But vaccination, when it arrived, was a great and unprecedented leap forward, destined to reach across the globe and become the most successful medical measure of all time.

Despite its enormous successes in the nineteenth century, for decades almost nothing was known about how smallpox vaccination worked, except for the fact that it did. There were two reasons for this. One was the mysterious nature of the agents of infection. It wasn't until the 1870s that an understanding of microbial pathogens first began to emerge. The other obstacle to understanding how vaccination worked had to do with the mysterious nature of immunity. Virtually nothing was known about how we naturally defend ourselves against infection until the discovery of antibody in 1890, which marked the birth of immunology.

In the last decade of the nineteenth century, news was breaking almost every month of new microorganisms, new diseases, new disease mechanisms; in the fledgling field of immunology, two passionate debates raged fiercely. The first was about whether inflammatory white cells were an abnormal, harmful consequence of disease or whether they were a defensive response to infection. The second was about which of the two phenomena associated with infection was important in immune defense: was it the invisible antibodies in blood and body fluids, or was it the circulating white cells whose behavior could be studied under the microscope?

Immunology in the first half of the twentieth century was dominated by the study of antibody, but the discovery of DNA in 1953 transformed immunological theories about how antibodies recognize germs. In 1958, an Australian scientist, Frank Macfarlane Burnet, solved the problem of how the immune system responds so quickly and with such exquisite specificity to so many different germs. His theory of adaptive immunity has stood the test of time.

Throughout the first half of the twentieth century, great progress in the art of vaccination was achieved not through advances in the field of immunology, but through progress in bacteriology, virology, and tissue culture. Vaccines against tuberculosis, whooping cough, diphtheria, and tetanus were introduced. By 1960, curative drug treatments for tuberculosis had become available, and the disease was preventable and controllable in many countries.

The tragedy of tuberculosis lay in the lingering decline of those infected, but polio was tragic for a different reason: the cruel suddenness with which it struck its victims down. The battle to defeat it unfolded not in Europe but in the United States in the first half of the twentieth century, when the fear of polio haunted every summer season.

Funded by the pioneering March of Dimes, the story of polio vaccines is one of rivalry, intrigue, and daring; it led, in the spring of 1954, to the great field trial of Jonas Salk's polio vaccine, which involved a half-million children. Despite a dreadful tragedy caused by contamination early in the trial, Salk's killed-virus vaccine was successful, and at last there was a weapon against this dreaded disease. But this was not enough to eradicate polio. Another, more potent vaccine was required, and through an extraordinary collaboration between American scientists and health authorities in the Soviet Union at the height of the Cold War, a live polio vaccine was delivered to the American people. By 1979 this vaccine had eradicated polio in North America.

In evolutionary terms, the adaptive immune system emerged about 400 million years ago. But throughout this time pathogens have also been evolving, and the ancient conflict between infectious agents and immunity works both ways. Viruses have evolved ingenious tricks to evade and subvert immune defense. Influenza does this every year by changing its coat. This was the problem faced by vaccinologists on the brink of the Second World War working to prevent another Spanish flu. This great pandemic killed 50 million people in 1918–1919, making it one of the deadliest disasters in human history. Today, a new flu vaccine must be manufactured every year to keep up with the changing virus, but many ingenious strategies are being used to make flu vaccination more successful.

In many ways the first vaccine was also one of the best. Jenner's discovery of a live vaccine that was safe in most people, and which produced immunity for many years with a single dose, remains hard to match. Jenner's vaccine was used throughout the World Health Organization (WHO) global campaign, which began in earnest in 1966, to eradicate smallpox. The campaign prevailed first in South America. Then the people of Indonesia's many islands were freed from the disease. Success in West Africa came next, followed by victory in India in the face of a smallpox epidemic. But Bangladesh proved even more dramatic: in a land beset by war, flood, and famine, smallpox was at last defeated by an army many thousands strong. The last-ever battle against natural smallpox was fought in the mountains and deserts and on the shores of East Africa, the cradle of humanity, where smallpox's final refuge was an ancient wandering people.

Shortly before the global eradication of smallpox, when success was already in sight, the WHO launched the Expanded Program on Immunization to build on their success. The vision was to make it possible for all children in all countries to benefit from life-saving vaccines against measles, polio, whooping cough, tetanus, diphtheria, and tuberculosis. Recently, new vaccines for hepatitis B, bacterial flu, streptococcal pneumonia, rotavirus, and rubella have been added. In the ever-flowing stream of human ingenuity that has its source in Jenner's discovery of smallpox vaccine in 1798, new advances are being made faster than ever before. At the frontiers of research, new principles and ever more complex strategies are being exploited to bring an end to global plagues. New vaccines for tuberculosis, HIV/AIDS, and malaria are in the pipeline, and some are already in clinical trials.

It is important not to underestimate the difficulties. And vaccination is by no means the only weapon needed. Sanitation and related public health measures, better hygiene, improved nutrition, extended health care infrastructure and health education, and the control of animal species that transmit human disease are all crucial elements in the fight against infection.

For smallpox, throughout the 179 years between the discovery of immunization and the last natural case, eradication often seemed an impossible goal. The successful smallpox campaign is reason for optimism. So too is the successful control of many deadly infections in developed countries. And the aspirations of developing nations to achieve the same freedom from disease as the developed world deserve all the resources and support that we can muster.

Effective eradication of epidemic disease does not mean the infectious agent disappears from the world. Species do, of course, become extinct, including microbial species, but smallpox is still with us. Official stocks of smallpox virus are maintained in American and Russian laboratories, and in all likelihood other states across the globe have acquired smallpox as a potential bioweapon. Should a universal flu vaccine be used across the world to bring an end to the problem of flu, the virus will remain in wild birds and mammals. When the global campaign to eradicate polio eventually succeeds, vaccine strains of poliovirus will continue to circulate among those who have been immunized and will need to be controlled. In this sense, effective eradication of infectious diseases may seem more attainable than the goal of pathogen extinction.

In 2012, as a result of the increasingly successful campaign to eradicate poliomyelitis, new cases of polio on the global scale reached an all-time low. Just 223 cases of paralysis were recorded. In Nigeria, Pakistan, and Afghanistan, the battle continues with renewed resolve. War, flood, famine, and outbreaks of disease will continue to be obstacles, as will cultural, political, and religious objections to vaccination. A new problem—acts of terrorism against unprotected vaccinators in remote and dangerous regions—may also remain with us. But these are the harsh realities of such a great endeavor. And for the worldwide campaign against polio, we stand on the brink of success.

1

THE POWER OF
THE INVISIBLE

"Of the causes of the Small-Pox; how it comes to pass that hardly anyone escapes the disease . . . Now the Small Pox arises when the blood putrefies and ferments, so that the superfluous vapours are thrown out of it, and it is changed from the blood of infants, which is like must, into the blood of young men, which is like wine perfectly ripened."

Abu Bakr Mohammad Ibn Zakariya Al Rhazi, ca. 910
Translated from Arabic by W. A. Greenhill, 1848

Imagine a great city at sunrise. The trains are running from the outer suburbs and small towns, carrying commuters to their daily work. The highways glitter with an endless stream of cars, mass-transit systems ferry hordes of people packed in tight, each wrapped up in their private world. Picture an air traveler circling high above, eager to be on the ground, peering through the aircraft window at the city spread below. Not one among its millions is visible to him. But of course, he has no trouble believing in the invisible.

It's simply a matter of scale. In every big city there are millions of bodies. And in every human body there are trillions of microbes, outnumbering the body's cells by ten to one. Most are beneficial and essential to health. But plenty are our enemies and many can be fatal. Fortunately for us, an army is mustered against them, constantly on call, and the streaming traffic of these inner forces is no less teeming than the rush-hour traffic in the most crowded cities on earth.

Cities favor the invisible enemy: it's so much easier for him to pass between one person and the next when people live in close proximity, sharing their resources and encountering hundreds of other individuals every day. But generally the interior army, protected behind barriers of skin and membrane, remains effective.

It's hard for us in the twenty-first century to imagine the lives of our ancestors before the age of modern medicine, sanitation, and nutrition. Disease shaped and altered human history, dramatically changing the course of wars, the scale of empires, and the fundamental texture of everyday life. In medieval times, values and beliefs were formed by the daily reality of death in the midst of life. Children were lucky to survive to adulthood; in Britain the average life expectancy was 30 years. Bubonic plague (the Black Death), arriving in Europe along the Silk Road from China, killed more than one-third of the European population between 1348 and 1350, precipitating a series of religious, social, and economic upheavals and hastening the end of feudal society.

Bubonic plague, which killed half its victims within days, is one of three terrible manifestations of the bacterium *Yersinia pestis*. Transmitted most frequently by rat-flea bites, it spreads through the lymphatics, blackening the grossly swollen lymph nodes in the groin and armpits as it overwhelms them. The name *bubonic plague* refers to these dark swellings, or *buboes*. But plague also shows itself in septicemic and pneumonic forms, the latter easily transmitted through airborne droplets. It was Daniel Defoe, merchant, writer, pamphleteer, and secret agent, who wrote the most compelling account of the Great Plague of London, which raged in 1665–1666. Ostensibly an artless, sometimes stumbling eyewitness account by an ordinary London trader, it is in fact a cleverly crafted novel written decades after the devastation.

> The face of London was now indeed strangely altered: I mean the whole mass of buildings, city, liberties, suburbs, Westminster, Southwark, and altogether; for as to the particular part called the city, or within the walls, that was not yet much infected. But in the whole the face of things, I say, was much altered; sorrow and sadness sat upon every face; and though some parts were not yet overwhelmed, yet all looked deeply concerned; and, as we saw it apparently coming on, so every one looked on himself and his family as in the utmost danger . . . London might well be said to be all in tears; . . . the shrieks of women and children at the windows and doors of their houses, where their dearest relations were perhaps dying, or just dead, were so frequent to be heard as we passed the streets, that it was enough to pierce the stoutest heart in the world to hear them. Tears and lamentations were seen almost in every house, especially in the first part of the visitation; for towards the latter end men's hearts were hardened, and death was so always before their eyes, that they did not so much concern themselves

for the loss of their friends, expecting that themselves should be summoned the next hour.

Close to where I live in Cambridge is a natural monument to the plague, which tells a more domestic tale. The village of Clopton once had a market, water mills, a flint church, and two moated manor houses. When the Black Death swept through England, wealthy survivors enclosed the abandoned land for grazing, which needed little man power (the decimated workforce was demanding higher wages). Today the village is a hollow in the empty hillside, full of willow trees and waving grass.

After the depredations of the fourteenth century, the Black Death returned to Europe in the great plagues of Seville, London, Vienna, and Marseille. But then something changed. After the middle of the seventeenth century, another dreaded infection replaced bubonic plague as the most feared affliction in Europe. Smallpox, the speckled monster, brought death and disfigurement on a massive scale, attacking the young in particular, indiscriminately ravaging not just the poor, but the richest and most powerful in every kingdom. It blinded many of its victims, killing a half million in a single year.

The burgeoning cities were particularly vulnerable. Tens of thousands died in the epidemics in Paris in 1723 and Rome in 1746 and 1754. The sickness was unimaginably worse than chickenpox—a disease that still infects children today. Ten to 30 percent of all infected people died. The death was horrible, prompting Thomas Macaulay, the British poet and historian, to call it "the most terrible of all the ministers of death."

What was the illness like? About ten days after infection, the patient, whether child or adult, was struck with a sudden fever, splitting headache and often backache, and sometimes vomiting. The rash appeared two to three days later, the fever subsided, and the patient, who had been feeling extremely ill, began to feel better. Over the next two weeks the fever returned and the rash began to change. The smallpox rash started in the mouth and throat and on the face; it then spread to the upper parts of the limbs and to the trunk and finally to the hands and feet. In the course of the infection it evolved from flat spots, which gradually became raised and grew hard, "like embedded lead shots." The spots then softened and began to fill, first with a clear fluid, then with pus. These pustules gradually flattened, dried, and scabbed. The fever ebbed and departed, and within about three weeks of first appearance the dried scabs fell off, leaving fibrous, pitted scars—the pockmarks—particularly on the face.

In mild cases the rash remained sparse and scarring was limited. But in moderate cases it covered the body, and in severe cases the pustules could be so crowded that they fused together in an intense rash called confluent smallpox with a mortality rate around 60 percent.

Smallpox is restricted to humans (no other species gets infected); patients are infectious for only a couple of weeks, and those who recover are immune to further attacks for the rest of their lives. This means that in small, isolated communities the disease can't persist and soon runs out of potential victims as immunity becomes established. In contrast, cities favor the infection. In large communities, even if they're isolated, the supply of vulnerable children is sufficient to maintain the infection, though the number affected might fluctuate wildly, peaking only once every few years with troughs in between until the birth of more susceptible children. In this situation the disease is called endemic, and the majority of its victims are young. In societies never exposed to smallpox (or where it has been absent for several generations), the whole population is at risk, and entire nations can be devastated. In 1521 the Spanish conquistador Hernando Cortés, with an army of only 900 men, defeated the entire Aztec nation of 5 million, largely because it had been ravaged by smallpox, which the Spaniards brought with them. Smallpox was a formidable enemy.

To the Aztec nation and the Inca nation, which fell to an even smaller army a decade later, it must have seemed that the gods were on the Spanish side: their own people were ravaged by disease while the Spaniards remained miraculously untouched. The Aztec emperor Cuitláhuac and the Inca emperor Huayna Capac died of smallpox carried by the invaders.

Endemic smallpox in Europe had long conferred protection through immunity on adventurers setting out to colonize new lands, but the Old World origins of the disease remain uncertain. We can't be sure if Old Testament accounts of plague and pestilence include descriptions of smallpox, and we don't know for certain if smallpox corresponds with accounts in early Chinese and Indian texts. The first definitive evidence of its existence comes from studies of Egyptian mummies, especially Ramses V who died in 1157 BC. His face, neck, shoulders, and lower abdomen are covered in pockmarks.

The plague of Athens in 430 BC may well have been smallpox, although measles and typhus are possible candidates as well. The outbreak began in the second year of the Peloponnesian War. It killed more than 30,000 people and reduced the population by 20 percent. Thucydides, an Athenian aristocrat, wrote a vivid account, describing the dead lying unburied, the temples full of corpses, and the violation of funeral rituals. Thucydides himself had the disease, but survived to write his celebrated history. He noted the nature of immunity: "the sick and the dying were tended by the pitying care of those who had recovered, because they knew the course of the disease and were themselves free from apprehensions. For no one was ever attacked a second time, or not with a fatal result."

In the first millennium AD, outbreaks of smallpox regularly struck the great civilizations of Asia and the Middle East, changing the course of conflicts and influencing the spread of Islam, Christianity, and Buddhism. The first detailed description of the illness was written in 910 by Al Rhazi, an eminent scholar in the early Islamic world and the first to distinguish between smallpox and measles. He proposed that natural exudation of superfluous infant humors was the cause of disease (since smallpox was endemic in tenth-century Baghdad, adults were immune and children the main victims). Another contemporary view was that the symptoms were due to the necessary expulsion through the skin of poisons acquired in the womb, consistent with beliefs about the unclean nature of menstrual blood. Al Rhazi's view of the infection was informed by his knowledge of classical medicine, and he advocated highly specific treatments to rebalance the humors: bleeding and cooling in the early stages of illness, and heat treatment (steam and swaddling) when the rash was erupting, together with dietary measures.

But there seems to be a mystery here. How could the most gifted physician of his age liken a terrible disease, which killed so many of its victims, to a natural process like the maturation of fine wine (as in the quote that opens this chapter)? The answer may well be that the disease afflicting tenth-century Baghdad was a distinct variant of smallpox caused by a virus with low virulence (now known as *Variola minor*). This would also explain why Al Rhazi ranked smallpox as less serious than measles. Whatever the reason, two centuries later virulent smallpox afflicted the crusaders. And when they returned to Europe they carried this prince of pestilence home with them.

2

CIRCASSIAN BEAUTIES AND PIONEERING WOMEN

"The smallpox was always present, filling the churchyards with corpses, tormenting with constant fears all whom it had stricken, leaving on those whose lives it spared the hideous traces of its power, turning the babe into a changeling at which the mother shuddered, and making the eyes and cheeks of the bighearted maiden objects of horror to the lover."

Thomas Macaulay, *The History of England, from the Accession of James II* (Volume 4), 1848

Six hundred years after Al Rhazi, despite the understanding of contagion and immunity (the knowledge that infections struck only once in the lifetime of an individual), imbalances in the four vital humors, as characterized by the Hippocratic physicians of ancient Greece, were still blamed for most illness. Blood, phlegm, black bile, and yellow bile were said to correspond with the four elements of air, water, earth, and fire. Rebalancing the humors remained the purpose of treatments such as cooling or heating the body, diet, exercise, herbal medicines, vomiting, purging, blistering, and bloodletting, right up to the nineteenth century. A particularly strange treatment for smallpox began with a Portuguese physician in the fifteenth century who recommended wrapping patients in red or purple cloth. The practice caught on and in the sixteenth century, when Queen Elizabeth had smallpox, she was wrapped in a red blanket.

Whatever the degree of ignorance and mystery surrounding the disease, one thing was certain: its sources were invisible. What sort of agent might cause such dreadful sickness?

We now know that virulent smallpox is caused by a virus called *Variola major,* a large, brick-shaped virus much bigger than flu but too small to be seen clearly under the light microscope. The virus is spread by airborne droplets from blisters in the nose and mouth of an infected person across distances up to two meters (and by direct contact with body fluids and contaminated objects such as bedding and clothes, which may harbor active virus for weeks or even months). However, the nature of viruses was not understood until the twentieth century. Modern genetic-sequence studies suggest that the virulent virus (*Variola major*) appeared through mutations of the milder form (*Variola minor*) between 400 and 1,600 years ago. This would explain the absence of smallpox from ancient texts and histories. *Variola minor,* in turn, was probably produced by mutation in a rodent virus more than 16,000 years ago.

Variola major was the principle form of smallpox throughout the eighteenth century, when the sickness ravaged Europe. The strategy of quarantine (begun in fourteenth-century Venice when plague-infected ships were isolated for 40 days—*quaranta* being Italian for *forty*) was practiced in smallpox outbreaks. But apart from that there seemed no way to check the dread disease. As the new century dawned, smallpox ended the Stuart royal line with the death of James II's grandson William in 1700 (his two sisters had already died of smallpox).

By the mid-eighteenth century, smallpox was a major endemic disease across the world. In Europe the death toll reached hundreds of thousands each year. At the end of the eighteenth century, about a third of all cases of blindness are thought to have been caused by smallpox. The disease checked armies, decimated populations, and ruined economies.

By this point in history, the Age of Science was underway, an age of visionaries and polymaths: Newton the physicist, mathematician, astronomer; Wren the architect and geometer; Hooke the inventor, biologist, and natural philosopher; Halley the astronomer and geophysicist. Seminal advances were made in the discovery of oxygen and hydrogen; the scientific design of roads, canals, and bridges; the mapping of the heavens; and harnessing the power of steam. But the practice of medicine remained more or less traditional, following tenets little changed over 2,000 years. In London most children in the eighteenth century had smallpox before they were seven, and little could be done to prevent it. In the single year of 1710, 3,000 people died from smallpox in London, and nine years later eminent physicians in the British capital were still arguing passionately about the merits of purging versus vomiting as treatments. When help did come, it was not from the medical profession.

As well as being the age of scientific visionaries, the eighteenth century was also a time of growing intellectual curiosity among more ordinary people who were privileged enough to indulge their new-found interests. This new class of intellectuals ranged from simple country doctors and clergymen to some of the highest in the land. Among them was an aristocratic Englishwoman who was to pioneer the first effective medical intervention against the scourge of smallpox. Born in London in 1689, her name was Lady Mary Wortley Montagu.

The Great Plague of London in 1665 and 1666, one of the last great visitations of the Black Death in Europe, was a disease of the common people. The English pamphleteer and author Roger L'Estrange said the heaviest impact was felt by the poor, "either crowded up into corners, and smothered for want of Aire; or are otherwise lost for want of Seasonable Attendance and Remedies." The rich were able to withdraw or keep themselves protected in the more salubrious parishes where rats and other noisome menaces were kept at bay.

But smallpox was a very different pestilence. When smallpox replaced plague as the most dreaded disease and the biggest killer, especially of the young, the patterns of affliction across Europe's social strata changed dramatically. The wealthy seemed just as susceptible as the poor, and the deaths of both children and adults born into the upper classes became a source of misery and despair that touched the highest families in the land. Smallpox killed Queen Mary II of England, Emperor Joseph I of Austria, King Luis I of Spain, Tsar Peter II of Russia, Queen Ulrika Eleonora of Sweden, and King Louis XV of France.

The family of Evelyn Pierrepont, Fifth Earl of Kingston-upon-Hull, was no exception: his son William died of smallpox at age 20, in 1713. Two years later, his daughter Mary, age 26, contracted the disease. Mary survived, but her looks were ruined. She had no eyelashes and her face bore deeply pitted scars. In the year of her recovery she wrote in verse:

How am I chang'd! alas! how am I grown
A frightful spectre, to myself unknown!

Her poem on the subject of smallpox ran to 96 lines.

Mary Pierrepont was a remarkable woman. She was adventurous and brave, and she pursued a wide range of interests, including poetry and women's rights, which she championed by example. Destined to marry the Honorable Clotworthy Skeffington, a man chosen by her father, she eloped instead with Edward Wortley Montagu. When Montagu was made British ambassador to the Sublime Porte of the Ottoman Empire, she joined him in Constantinople. Her letters home provide a vivid picture of life in the Muslim Orient and became an inspiration to later generations of women who wrote about their travels. Lady Mary had no great opinion of the writers of her time, deriding the quality of European travel literature as nothing more than "trite observations" written by boys who "only remember the best wine or the prettyest women." She learned the language and affected Turkish dress, donning a turban and full regalia, "the very full drawers of which," she wrote to her sister, "reach to my shoes and conceal the legs more modestly than your petticoats." Her collected letters include an increasingly caustic feud with the poet Alexander Pope, once a friend and admirer, in which she mercilessly parodied his work. Because of her defiant attitude toward convention and her passion for Turkish life, Lady Mary became interested in a traditional practice widespread among the Greek community in Constantinople.

In Europe at that time, there was a fascination for the Orient that included tales of the women of Circassia in the North Caucasus, now part of modern Russia. The women of this mountainous region were often characterized as the ideal in feminine beauty in European literature and art. The loveliest women in the harem of the Sultan were said to have arrived across the Black Sea from Circassia, where their families ensured their beauty by protecting them against smallpox. This they did by an ancient practice, which came to be known as variolation (*variola* is another name for smallpox from the Latin *varius*, meaning spotted or variegated). The tradition was to take material from the rash of someone with a mild case of smallpox and transfer it, in very small amounts, into the body of a healthy infant to produce a mild and limited infection and confer lifelong protection.

While this tradition may well have preserved the beauty of Circassian girls, the origins of variolation are far more widespread than this romantic tale portrays. The practice was used in ancient China, where the custom, according to Chinese texts, was to dry the smallpox material and blow it into the nose of the recipient through a silver tube—the right nostril for a boy, the left for a girl. Another method, more common outside China, was to make a small, shallow incision in the arm and place smallpox material under the skin with a needle. In this way the infectious material from mild cases was introduced in small amounts via an unnatural route, where it remained confined in damaged tissue with little access to the circulation. The result was usually a limited illness from which the recipient could readily recover without facial scarring. It seems, as well, that variolation was practiced in certain African countries and in India for centuries. Less invasive, and therefore less effective, practices were also a part of rustic traditions in Europe. In one version, smallpox blister fluid was rubbed onto punctured or abraded skin; in another, dried smallpox scabs were clutched in the hand. But because these obscure traditions were confined to simple country folk, they rarely entered the consciousness of educated people, and if they did they were regarded as ignorant superstition. Lady Mary was enthusiastic about her newfound knowledge. In April 1717 she wrote to her friend Sarah Chiswell, "A propos of Distempers, I am going to tell you a thing that I am sure will make you wish your selfe here. The Small Pox so fatal and so general amongst us is here entirely harmless by the invention of engrafting (which is the term they give it). There is a set of old Women who make it their business to perform the Operation." True to her words, Lady Mary proceeded to instruct the Scottish surgeon at the British Embassy in Constantinople, Charles Maitland, to variolate her six-year-old son Edward. The variolation was performed in March 1718. Maitland later wrote that

> she [Lady Mary] first order'd me to find out a fit Subject to take the matter from: and then sent for an old Greek woman, who had practis'd this Way a great many Years: After a good deal of Trouble and Pains, I found a proper Subject [smallpox victim], and then the good Woman went to work; but so awkwardly by the shaking of her Hand, and put the Child to so much torture with her blunt and rusty Needle, that I pitied his Cries . . . and therefore inoculated the other arm with my own instrument, and with so little Pain to him that he did not in the least complain of it. The Operation took in both Arms, and succeeded perfectly well. After the third day, bright red Spots appeared in his Face, then disappear'd . . . betwixt the Seventh and Eighth Day . . . the Smallpox came out fair . . . the young Gentleman was quickly in a condition to go Abroad with safety.

The term *inoculation* (a gardening term, meaning ineyeing a bud on another plant) was also used to describe variolation. Edward, like his mother, proved to be adventurous and independent. At fifteen he ran away to sea, and when he grew tired of his disguise he owned up to the captain, using his inoculation scar as proof of his identity.

Not long after Lady Mary's return to London, an epidemic of smallpox broke out, and she began a spirited campaign to promote the practice of variolation to the establishment. In 1721 she persuaded Charles Maitland (now retired to England) to inoculate her three-year-old daughter Mary; this was the first known immunization by a medical professional performed in England. The event received considerable publicity. Maitland insisted on spreading the great responsibility of the operation by having two physicians as witnesses, and several other eminent men came to inspect the little girl's pocks. After a slight fever Mary recovered well. She later married the Scottish nobleman Lord Bute (prime minister in 1762) and had eleven children.

The British smallpox epidemic of 1721 aroused great interest among the aristocracy as to what could possibly be done. In particular the Princess of Wales, Caroline of Ansbach, fearful for her children's lives, became intensely interested in the possibilities of variolation, perhaps through Lady Mary's efforts but also through Sir Hans Sloane, court physician to King George I. Charles Maitland himself also began to publicize the practice, keen to prove the value of the new procedure. As a result of Princess Caroline's concerns, the king was petitioned by his physicians, who hoped to establish the safety of the strategy so it could be used to protect the royal children. After due consideration and legal consultation, the king agreed to their proposals.

At Newgate Prison on the morning of August 9, 1721, three male and three female prisoners who had been condemned to death became the subjects of the "Royal Experiment." The king had agreed that, should they survive the hazards of the operation, the condemned prisoners would receive the royal pardon and go free.

No doubt the six prisoners were more than ready to be experimented on—the infamous prison was a forbidding, overcrowded place, and in the poor sanitary conditions disease was rife. Death rates were high, especially in winter. In 1726 jail fever killed 83 prisoners, and in 1750 the prisoners brought the fever into the Old Bailey courtroom. The Lord Mayor Sir Samuel Pennant, two judges, an alderman, an under sheriff, and 50 court officials died of the infection. Jail fever is now known to be typhus, a bacterial disease transmitted, like bubonic plague, through the bites of lice and fleas. It was often confused with typhoid, a waterborne infection. The name *typhus* comes from a Greek word meaning smoky or hazy, said to describe the minds of its victims. Between 1755 and 1765, 132 prisoners died. Escapes by desperate, overcrowded prisoners were frequent.

As the sun rose over Newgate in August 1721, six prisoners were ready to escape it, and escape too the long cart-ride to the gallows. Charles Maitland performed the inoculation procedure, watched by an audience of 25 eminent and learned men including Sir Hans Sloane, members of the College of Physicians, and fellows of the Royal Society. The unprecedented events were widely reported in the press. With the eyes of the London medical establishment upon him, Maitland made small incisions on the right arms and legs of five of the six prisoners and carefully inserted material taken from the skin of a smallpox patient. Three days later he examined the inoculation sites and decided they were not as inflamed as he'd hoped. Anxious to succeed, he obtained fresh smallpox material and repeated the insertion. But according to Sir Hans

Sloane's account written in 1736 (but not published until 1756), one prisoner was treated differently: "Upon my speaking to Mr. Maitland, he undertook the operation, which succeeded in all but one, who had the matter of the small-pox put up her nose, which produced no distemper, but gave great uneasiness to the poor woman."

The next day smallpox symptoms appeared in all the prisoners except the woman inoculated through her nose. After a brief illness each of the subjects began to feel better. Interest in the fortunes of Maitland's patients was high, and they were visited every day by physicians and other dignitaries while the press reported their progress. All six recovered quickly, received pardons from King George and his council, and were freed from Newgate Prison on September 6. In the *Philosophical Transactions of the Royal Society,* volume 49, Sir Hans Sloane describes what happened next.

> After their recovery, in order to obviate the objection made by enemies of the practice, that the distemper produced by it was only chicken-pox, swine-pox, or the petite *varole volagere,* which did not secure persons against having the true small-pox, Dr. Steigertahl, physician to the late king, and I joined our purses to pay one of those, who had it by inoculation at Newgate [a girl of nineteen], who was sent to Hertford, where the disease in the natural way was epidemical and very mortal and where this person nursed, and lay in bed with one, who had it [a ten year old boy], without receiving any new infection.

Princess Caroline was impressed and reassured but not entirely convinced. After all, the lives of her five small children were at stake, and the experiment had been on adults. Maitland performed several more adult variolations, but in mid-November the London newspapers announced that the Princess of Wales had drawn up a list of all orphans in St. James Parish, Westminster, who had not yet had smallpox. She intended to have them inoculated at her own expense, though one resourceful girl would get the better of her. According to Sloane's account of 1756:

> To make further trial, the late Queen Caroline produced half a dozen of the charity children belonging to St. James's parish, who were inoculated, and all of them, except one, who had had the small-pox before, tho' she pretended not for the sake of the reward, went thro' it with the symptoms of a favourable kind of that distemper.
>
> Upon these trials, and several others in private families, the late queen, then princess of Wales, (who with the king always took most extraordinary, exemplary, prudent and wise care of the health and education of their children) sent for me to ask my opinion of the princesses.

I told her royal highness, that, by what appeared in the several essays, it seemed to be a method to secure people from the great dangers attending that distemper in the natural way . . . but that not being certain of the consequences, which might happen, I would not persuade nor advise the making of trials upon patients of such importance to the publick. The princess then asked me if I would dissuade her: to which I made answer that I would not, in a matter so likely to be of such advantage.

The "Royal Experiment" (and other private inoculations) had dramatic consequences. Several prominent political figures called on Maitland to inoculate their own children. The royal initiative was crucial in fueling much wider interest. The momentum increased. Caroline requested permission from the king to have her daughters variolated. The king sought advice from Sloane, who later wrote, "I told his majesty my opinion that, it was impossible to be certain that raising such a commotion in the blood, there might happen dangerous accidents not foreseen." King George, asserting that all medical procedures carried risk, granted Caroline her wish. On April 17 eleven-year-old Princess Amelia and nine-year-old Princess Caroline were inoculated by the Royal Surgeon Claude Amyand with the assistance of Maitland.

Caroline was not just a privileged mother concerned for her children; she was a well-informed intellectual with a serious interest in science. She fostered debate about the natural philosophies of Newton and Gottfried Leibnitz, the German mathematician, with whom she corresponded as a friend. Her daughters' variolations were successful, and her example led to unprecedented interest from the rich and privileged in the great possibilities of inoculation.

The new practice was very much in need of highborn champions. Variolation was in fact already known to British scientists and physicians, but nothing had been done about it. In 1700 Chinese variolation had been described to the Royal Society by a member of the East India Company. Fourteen years later two respected Italian physicians working in the Orient, Dr. Emanuele Timoni and Dr. Jacobo Pylarini, sent detailed reports, which the society duly published.

Crucial information came too from the New World, where the secrets of variolation were revealed in 1706 by a North African man transported to New England as a slave. We do not know his name, but his owner, the puritan minister Cotton Mather, named him Onesimus after a biblical slave converted to Christianity. Onesimus told his master about the practice of variolation he had known as a child in Africa. Boston was plagued by repeated smallpox outbreaks, and Mather, who had also read the Italian reports on variolation published by the Royal

Society, responded by writing enthusiastically to influential physicians in London. Yet in England no one wanted to risk their reputation or take on the difficult task of promoting this exotic custom to the medical establishment. Nothing came of the knowledge until Lady Mary Wortley Montagu and Princess Caroline took up the challenge.

In Britain and in Germany, inoculation became quite a familiar procedure among the aristocracy, but in America it was less confined to the privileged few. In the Boston smallpox outbreak of 1721, Cotton Mather, enlightened by Onesimus, wrote to all the doctors of the town appealing for a trial of the exotic practice. Only one, Dr. Zabdiel Boylston, responded. Boylston, a careful and highly respected practitioner, tried the procedure on his only son and two slaves: a man and a boy. All three recovered in about a week. Dr. Boylston went on to inoculate 242 people during the Boston epidemic.

Despite its success in both Europe and America, inoculation had many passionate opponents. The British surgeon Legard Sparham wrote a book, saying that he "never dreamt that Mankind would industriously plot to their own ruin, and barter Health for Disease." The Reverend Mr. Massey, preaching in St Andrew's, Holborn, claimed that the practice tended to "banish providence out of the World and promote the encrease of Vice and Immorality." In America, ferocious criticism of Boylston and Mather came from both religious and medical groups. Mather recorded in his diary that his efforts against smallpox had "raised a horrid Clamour," and that he and Boylston had become the objects of "their furious Obloquies and Invectives." There was, in fact, some substance to the criticisms: as variolation became more widespread and information on its outcome continued to accumulate, it became clear that variolation did cause some deaths, and that without proper isolation, recipients of variolation could infect other individuals via the natural route, leading to the spread of serious disease.

When we contemplate the value of variolation from our modern standpoint, it's hard for us to grasp the realities of life in the eighteenth century. For modern medicines that we give to healthy people, our concerns focus on extremely small risks of hundreds of thousands to one. Live polio vaccine, for example, at the high end of vaccine risk, has, at worst, a global average chance of causing paralysis in one in 750,000 recipients. The mortality rate for smallpox was extremely high: 1 to 3 patients in every 10 died, and disfigurement at least was the expected outcome. As larger numbers of people were variolated, what we would now call the risk-to-benefit ratio of the procedure slowly began to emerge. In the seven years following Princess Caroline's example, 897 variolations were recorded in Britain, America, and Hanover, and 17 of the recipients died of the resulting disease. In other words 2 percent (compared with an average 20 percent in natural infection). This means that anyone contemplating variolation at this time had to face a 1 in 50 chance of death. But in 1729, since the great majority of children were likely to get smallpox, with a 1 in 5 chance of death, the measure made good sense.

But the French did not agree. The medical establishment, embodied in the Faculté de Médicine of the University of Paris, rose up against variolation, condemning it as a "useless, uncertain and dangerous practice." As a result, the adoption of variolation was delayed in France by about 40 years. Voltaire's was a rare voice in favor. In his *Eleventh Letter on the English*, referring to Lady Mary Wortley Montagu, he wrote:

> Had the lady of some French ambassador brought this secret from Constantinople to Paris, the nation would have been for ever obliged to her. Then the Duke de Villequier, father to the Duke d'Aumont, who enjoys the most vigorous constitution, and is the healthiest man in France, would not have been cut off in the flower of his age. The Prince of Soubise, happy in the finest flush of health, would not have been snatched away at five-and-twenty, nor the Dauphin, grandfather to Louis XV, have been laid in his grave in his fiftieth year. Twenty thousand persons whom the small-pox swept away at Paris in 1723 would have been alive at this time.

Understandably, in the decades that followed, interest in variolation waxed and waned in parallel with the comings and goings of the "speckled monster." When smallpox returned to London in 1746, the Smallpox and Inoculation Hospital was founded to give free care to patients and make variolation available to the poor. The practice was also extended to the countryside. In 1757 an eight-year-old orphan was variolated near the town of Gloucester, and we have a graphic description of this particular event. The process was elaborate and very unpleasant because the medical establishment, unwilling to relinquish tradition and keen to surround the procedure with medical mystique, had added a period of fastings and bleedings for six weeks before the inoculation. The boy was "bled to ascertain whether his blood was fine; was purged repeatedly, till he became emaciated and feeble; was kept on very low diet, small in quantity and dosed with a diet-drink to sweeten the blood. After this barbarism of human-veterinary practice he was removed to one of the inoculation stables and haltered up with others . . . By good fortune [he] escaped with a mild exhibition of the disease."

In time, these cruel and unnecessary additions to the practice of variolation began to be reduced. A practitioner named Robert Sutton, together with his sons, refined the practice, making substantial improvements that reduced and then eliminated the barbaric preparation and used material from early (day four) rashes, which produced a much milder illness. One son, Daniel Sutton established his own practice and eventually claimed to have inoculated 40,000 people with only five deaths, though passing on the induced infection remained a significant danger. Other innovators in America and Europe focused on reducing the dose. When Catherine the Great sought variolation, she asked the Russian ambassador to London to find her the best practitioner. Daniel Sutton declined, but an adherent, Thomas Dimsdale, agreed and traveled in secret to perform the operation. When all turned out well and Catherine began to encourage the practice in Russia, Voltaire wrote to thank her for her lesson to the "ridiculous Sorbonne and to the argumentative charlatans in our medical schools! You have been inoculated with less fuss than a nun taking an enema."

After helping to set up variolation hospitals in St. Petersburg and Moscow, Dimsdale received a fairy-tale reward including a fee of £12,000, a pension of £500 a year, diamond-studded portraits of Catherine and the Grand Duke, and the hereditary title "Baron of the Russian Empire."

The French eventually embraced variolation, convinced by the efficacy and relative safety of the Sutton method, which had reduced deaths to about one in a thousand; isolation was rigidly enforced to prevent contacts being infected by the natural route, which often led to serious illness. In continental Europe, celebrated recipients of variolation included the royal children of the empress of Austria and members of the French royal family. Following the death of Louis XV from smallpox, Louis XVI was variolated in 1774.

Accounts of individual variolations are rare, and we are fortunate to have a full description of the Gloucestershire orphan and what he had to go through in 1757, before the procedure was refined. The account was written by the Reverend Thomas Dudley Fosbroke, the curate of the parish of Horsely in Gloucestershire at the end of the eighteenth century. As to the unfortunate orphan, his name was Edward Jenner.

3

THE MAKING
OF JENNER

*"I lovd the pasture with its rushes & thistles & sheep tracks I adored
the wild marshy fen with its solitary hernshaw sweeing along in its
mellancholy sky I wandered the heath in raptures among the rabbit burrows & golden blossomd furze I dropt down on the thymy
molehill or mossy eminence to survey the summer landscape as full
of rapture as now."*

John Clare, c. 1821

Edward Jenner was born on May 17, 1749, in the country
town of Berkeley in the southern reaches of the Severn Vale,
a hundred miles from London. His father, Stephen, was vicar
of Berkeley, a prosperous and peaceful living in which he enjoyed the
patronage of the Earl of Berkeley and his family. Edward was the eighth
child, six of which would survive to adulthood. When he was five, his
mother, Sarah, died soon after giving birth to a son who lived a single
day. Two months later his father died, leaving Edward in the care of his
sisters Mary, Sarah, and Anne. Mary, the oldest, and a favorite aunt,
Deborah, took the place of his mother. His brothers, Stephen and Henry,
students at Oxford at this time, became his official guardians.

The quiet countryside of river, wood, and meadow around his home,
with its abundant natural life, became a fascination for the young Edward. In 1757, at age eight, he was sent as a boarder to Wotton Under
Edge grammar school, six miles from Berkeley. Here he began a collection of dormouse nests that he discovered in the ancient woods of oak
and ash nearby. It was here too that he suffered the severe form of variolation when smallpox threatened the little market town. A year later he

was sent to grammar school at Cirencester, a much larger community, where he studied under the Reverend Dr. Washbourne and became interested in fossils, which he searched out on his endless country rambles. Jenner did not take too well to Greek and Latin, preferring instead the study of nature. He made lifelong friends with other boys who grew up to be naturalists and prominent medical men. Perhaps because his brothers had become churchmen, and because of the expense of an Oxford education to a fatherless family, Jenner set out on a different path. At thirteen he was apprenticed to a local surgeon, Daniel Ludlow of Chipping Sodbury near Bath, and toward the end of his six-year training, at age nineteen, he learned about an uncommon distemper of cattle called cowpox that could infect milkmaids.

Six years was sufficient training at that time to set up as a country surgeon (without a medical diploma), but Daniel Ludlow had trained at St. George's Hospital London and this may have influenced Jenner's next move. In 1770 he took lodgings in London in the house of John Hunter, an eminent Scottish surgeon, on Jermyn Street. Here, for two years, he studied surgery with Hunter at St. George's Hospital, and anatomy, physiology, and midwifery with Hunter's elder brother, William, at the new Anatomy School on Windmill Street. The brothers championed accurate instruction using fresh human specimens, often exhumed for the purpose, rather than the traditional stylized drawings of the human body. For board and tuition Jenner paid £100 a year.

Only Oxford and Cambridge could confer official medical diplomas at that time, but Jenner at least gained certificates of attendance and became an expert in the practice of medicine and surgery. He made friends with two fellow student-boarders, Henry Cline and Everard Home, future presidents of the Royal College of Surgeons.

John Hunter was an excellent mentor for the young Jenner. Hunter had an endless fascination for the natural world and kept a small private zoo at Earl's Court outside the city where he collected such curiosities as leopards, ostriches, buffaloes, and jackals; he dissected other specimens taken from the menagerie at the Tower of London. Hunter also made extensive studies of animal behavior and other topics beyond surgery. It was at this time that Captain Cook returned from his first voyage of scientific exploration, and the naturalist of the expedition, Joseph Banks, at Hunter's suggestion, asked Jenner to catalog his collection of botanical specimens and other treasures. (Banks eventually had more than 80 plant species named after him.) Jenner accomplished the task so well that Banks invited him on his next voyage with Cook. Jenner, however, declined the opportunity, preferring the tranquility of his country practice to adventure on the high seas.

It was in 1772, at age 23, that Jenner returned to his native countryside, setting up house in Berkeley with his churchman brother Stephen. Here he pursued the pleasant life of a cultivated country doctor. John Hunter's wife, Anne, a poet and musician and a friend of Joseph Haydn's, had influenced the impressionable Jenner, who enjoyed singing, playing the violin and flute, and writing poetry. His friend Edward Gardner provides a picture of Jenner around this time, remarking on his fine blue coat with yellow buttons, his buckskins, his well-polished jockey boots with handsome silver spurs, his smart whip with a silver handle, and his hair done up in the latest fashion.

Jenner's poetry was another aspect of his thoughtful reflections on the natural world. His playful poem "Signs of Rain" describes how

Last night the sun went pale to bed,
The moon in halos hid her head . . .

Despite his relative seclusion, Jenner's intellectual life remained lively. He joined a society, the Convivio Medical Society, founded by the country surgeon John Fewster, which met each week to dine, discuss cases, and debate any subject of interest under the sun. The meetings were held at the Ship Inn in Alveston (still a thriving establishment today), about ten miles from Bristol. Jenner's broad interests eventually extended to the fledgling art of aeronautics. In 1784, less than a year after the Montgolfier brothers' hot-air balloon ascent, Jenner and Lord Berkeley organized an unmanned flight by a hydrogen-filled balloon. This pioneering craft traveled a distance of ten miles. Jenner pursued it on horseback, little knowing he was riding toward his own destiny. For where it fell to earth (drawing a considerable crowd), he met his future wife, Catherine Kingscote.

A year after this daring aerial experiment (producing hydrogen with iron filings and sulfuric acid was a very dangerous procedure), Jenner bought a home in Berkeley: an elegant Queen Anne house close to the imposing church where his father had been vicar for 29 years. He paid £600 for it. Much of what we know of Jenner's life comes from the writings of his friends, especially his biographer, the physician John Baron, whose somewhat adulatory account was published in 1838, sixteen years after Jenner's death. Jenner also maintained a lively correspondence with John Hunter, who remained his friend and mentor and whose appetite for biological specimens (dead or alive) seemed to be insatiable. The following (reproduced by the Royal College of Surgeons) is typical. Judging by Hunter's diplomatic postscript, it seems one live specimen was lucky to survive.

Dear Jenner

I rec'd yours by Dr Hicks with the Hedge Hog alive, I put it in my garden but I want more. I will send you the picture but by what carriage or by what place. I have a picture of Barret & Stubbs. The Landscape by Barret, and a Horse frightened at the first seeing of a lion by Stubbs. I got it for five guineas. will you have it? I have a dearer one, and no use for two of the same master's. but do not have it excepting you would like it for I can get my money for it. I am glad you have got Black birds nests. Let me know the expense you are at for I do not mean that the picture is to go for any of it only for your trouble

Ever yours
JOHN HUNTER
N.B. I should suppose that Hedge Hogs would come in a Box full of holes all round fill'd with Haye, and some fresh meat put into it.

We also know from this correspondence that before meeting Catherine, Jenner had been disappointed in love. Hunter wrote to console Jenner, suggesting sympathetically that he recover his spirits by immersing himself in the study of hedgehogs (obvious, really, when you think about it). Whether or not in immediate response, Jenner pursued a detailed study of hedgehog behavior, measuring temperatures and studying stomach contents. When spring came to the Severn Vale, he did the same with birds, noting, in contrast to the hibernating hedgehogs, their constant weight and full stomachs containing exotic seeds—important evidence for bird migration at a time when swallows and swifts were thought to overwinter in the mud of England's ponds and streams. Jenner's most extraordinary contribution to Hunter's specimen collection was a whale. The five-meter-long bottlenose whale was shot in the Severn Estuary not far from Berkeley. Jenner ensured the poor creature ended-up in Hunter's museum.

At Hunter's suggestion, Jenner also took up the study of cuckoos. Bringing his exceptional powers of observation and persistence, he first noted that the female cuckoo tends to choose species whose eggs resemble her own in color. He confirmed that the incubation period was shorter than the host species so the baby cuckoo was the first to hatch. He wrote a scientific paper (accepted by the Royal Society) in which he speculated that the duped parents abandon their offspring in favor of the endlessly demanding giant in the nest.

When spring returned to Berkeley Vale, Jenner resumed his studies, carefully placing cuckoo eggs in nests convenient for study. This enabled him to observe the remarkable behavior of the newly emerged cuckoo as it maneuvered the other eggs in the nest into the special hollow on its back and hoisted them out. He saw young cuckoos do the same with helpless chicks; anxious to confirm the instinctive drive behind this strange behavior, he planted two cuckoo eggs in the same nest. As soon as they hatched, he was able to witness a contest in which each chick got the upper hand in turn, until one of them succeeded in tipping its rival over the side. To broaden and verify his observations, he established a network of correspondents to collect information on his behalf. His nephew Henry, an apprentice surgeon in his care, helped him, traveling long miles on foot and reporting back on nests ripe for study. Somehow Jenner managed to retrieve his original paper on the cuckoo and rewrite it to describe what really happens. He formally submitted his study (performed in "some of my leisure hours") as a letter to John Hunter, who was a fellow of the Royal Society, and it was published in the society's journal in 1788. The following year, supported by his influential friends, he was made a Fellow of the Royal Society at the age of 40.

In March 1788, in a small upland church on the southwest edge of the Cotswold Hills, Jenner married his sweetheart, Catherine Kingscote. He was 38 and she was 27. Kingscote Church contains a memorial plaque to the wedding, which tells us that "his work in connection with the introduction of vaccination has made him for ever dear to the human race," and adds, "His marriage brought him much happiness."

Three years later, after twenty years as a general practitioner and surgeon, Jenner obtained a medical diploma through nomination from the University of St. Andrews, which enabled him to focus on working as a physician. He established a practice in the fashionable spa of Cheltenham, which he kept up each summer season until 1815, receiving many referrals from John Hunter's former students in London.

Unsurprisingly perhaps, Jenner didn't let his new status affect his concern for poorer patients and what might be done for them. Together with his friends he founded a new society, the Medico Convivial. It quickly became the Medical Society for Gloucestershire and included representatives from Bath, Bristol, and Hereford who traveled long distances in the saddle or by horse-drawn gig.

One of the subjects that came up for discussion was angina, a complaint of unknown cause. Jenner had the opportunity to perform autopsies on two people who had died of the affliction, and he painstakingly observed the narrowing and calcification (hardening) of the coronary arteries supplying blood to the heart. Since this was the only abnormality, he concluded (correctly) that it was the cause. In 1799 his friend Caleb Parry published a book on the condition, crediting Jenner with the discovery. When John Hunter fell ill in 1785, Jenner wrote to his doctor William Herberden (the first man to recognize angina as a condition) and suggested angina was the culprit, but characteristically entreated him not to let Hunter know of this serious possibility. When Hunter died eight years later, an autopsy revealed obstruction of the coronary artery.

Jenner's strong preference for a quiet country life is perhaps understandable given the times in which he lived. While he studied birds and hedgehogs, discovered the cause of angina, and improved the purity of certain medicines, the bloody battles of the American War of Independence were raging. Among the great forces shaping events, smallpox was ever present. George Washington called it "more destructive to an army in the natural way than the sword." Washington's fear of smallpox in Boston dictated his strategy, preventing him from attacking the city for nine months. When the British left Boston at last, he sent in a force of 1,000 men known to have had the disease. As the displaced New Englanders returned, the Boston smallpox epidemic spread. For a short time the practice of variolation was encouraged, first for vulnerable troops and later for the general population, though failure to isolate inoculees was an ever-present danger. Those in contact with variolated individuals ran the risk of infection through the natural route, leading to serious disease. Social gatherings for variolation became fashionable, and John Hancock, Bostonian merchant and political leader, wrote to George Washington that his wife "would esteem it to have Mrs. Washington take the smallpox in her house."

Although Washington was against variolation (because without proper isolation it spread disease), the decimation of troops and the failure of recruitment through fear of smallpox persuaded him to change his mind. In 1777 he wrote, "Finding the smallpox to be spreading much, and fearing that no precaution can prevent it from running through the whole of our Army, I have determined that the troops should be inoculated." It was an extraordinarily difficult thing to undertake, and he was forced to proceed in secret to avoid the British attacking a temporarily weakened force.

In the War of Independence, black troops suffered most from the ravages of smallpox, often because they were confined in overcrowded and insanitary camps and shipboard quarters. In Quebec in 1776 smallpox defeated the besieging colonial force when half their numbers were infected. A future president, John Adams, called the epidemic "ten times more terrible than the British, Canadians and Indians together. This was the cause of our precipitate retreat from Quebec." Some claim that smallpox is the reason Canada remains today a part of the British Commonwealth.

4

WHY NOT TRY THE EXPERIMENT?

"I profess to learn and to teach anatomy not from books but from dissections, not from the tenets of Philosophers but from the fabric of Nature."

William Harvey, 1628

Not half as famous as Newton's apple, less revered than Florence Nightingale's lamp, the story of the first vaccination still flickers in the margins of collective memory: the milkmaid, the small boy, the surgeon with his lancet poised. But the true context in which Edward Jenner performed his historic experiment, and his reasons for undertaking it, are often little understood.

Jenner's daily practice as a country doctor included surgery, general medicine, midwifery, diagnosis, and dispensing. He made his rounds on horseback, setting out from Chantry Cottage in the shadow of Berkeley Castle and riding through the country lanes of Gloucestershire. Berkeley Vale lies wide and open to the sky; unsheltered in winter, transformed in summer by a host of wayside flowers: willow herb and meadowsweet, cranes bill and pennycress, wild mint and basil. To the west the vale is bordered by the broad reaches of the Severn and the rising Forest of Dean. To the east, the Cotswold escarpment springs up from the level meads to a rolling upland—a place of sudden hills and hanging woods, secret dells and sloping meadows curtained round with trees. The hills command dramatic views of the great, unbridged river winding through its valley and the mountains of Wales beyond.

As a surgeon, Jenner was highly respected. Once, when an urgent and difficult case presented itself in Gloucester Hospital, Jenner was hastily summoned (eighteen miles on horseback). He performed an emergency operation on a strangulated hernia that would certainly have been fatal; he succeeded so well that the patient lived happily for another decade. Jenner was often in demand as a consultant in neighboring practices and had to travel long distances from home. In 1775 John Hunter offered him a partnership to teach natural history and comparative anatomy in London (Jenner would have had to invest £1,000 in the enterprise), but he decided against it, preferring the country life.

One of his medical duties, when a smallpox epidemic threatened the locality, was to protect by variolation those most at risk. In 1789 he variolated his own son, Edward, and two servants when smallpox threatened the household. Jenner used the improved Sutton method, which had a relatively small chance of causing serious illness, and in keeping with his generous character, he extended the inoculations to ordinary working folk in his care, not minding that they couldn't pay. This meant that a new class of ordinary people was added to Jenner's experience of inoculation.

In best practice, conscientious variolaters repeated the injection of smallpox material a few weeks later to confirm that the inoculation had worked, the second injection producing only a local reaction and no further malaise. When milkmaids were the subjects of variolation, Jenner observed that the initial procedure often produced not the expected illness but just a local blister, as if these particular patients were already protected. This set him thinking.

From quite early on, Jenner would have been familiar with the folk wisdom telling that milkmaids didn't get smallpox. The fair, unblemished country maiden was a favorite ideal of the time. Vermeer's *The Milkmaid* is a famous instance, but Jean-Baptiste Greuze's *La Laitière,* with her perfect milk-white skin and softly blushing cheeks, is a better example. According to John Baron (Jenner's biographer), while Jenner was still an apprentice a country girl told him, "I cannot take that disease [smallpox] for I have had the Cow Pox." The folklore on cowpox as protection against smallpox, however, was not taken seriously by most medical men. In 1754, the Swiss physician Theodore Tronchin, a highly influential doctor and proponent of variolation (he'd inoculated several thousand in Europe), heard of the claim that Gloucestershire milkmaids didn't get smallpox and dismissed it as nothing more than ignorant superstition.

Despite this, Jenner's friend and colleague John Fewster had written a paper for the London Medical Society in 1765 on "Cow pox and its ability to prevent Smallpox"; the topic was discussed at Fewster's Convivio Medical Society where Jenner was an enthusiastic participant. But

Fewster concluded that since inoculation worked reasonably well there was no need to try and improve on it.

As he reflected on the stories and the evidence he was gradually collecting, Jenner would doubtless have in mind the advice of his mentor John Hunter regarding hedgehog hibernation, cuckoo behavior, and many other topics: "Why think? Why not try the Experiment?" A century earlier William Harvey (who discovered the circulation of the blood "by reason and experiment") held the same view, urging his colleagues "to search and study out the secrets of nature by way of experiment." Hunter and his pupil Jenner wholeheartedly agreed.

Jenner, always impressive in his powers of observation, was able to distinguish two forms of cowpox by the appearance of the pustules. One, pseudocowpox, was (and remains) a common infection of the udder of dairy cows, often a chronic grumbling problem with most of the herd becoming infected. In contrast, true cowpox was a relatively uncommon infection. As Jenner rightly believed, only true cowpox was capable of protecting against smallpox. By 1788 he had made highly accurate sketches of cowpox pustules and taken them to London to discuss with his colleagues. Jenner, though, was easily diverted by his many other interests and enjoyed plenty of conjecture time before he finally took the plunge.

In 1795 Jenner suffered a serious attack of typhoid fever, which almost killed him. Typhoid, named for its resemblance to typhus, is caused by an unrelated bacterium, *Salmonella typhi,* which is transmitted in drinking water contaminated with human waste or fecal contamination of food. In Jenner's day the disease was attributed to noxious miasmas rising from putrid matter (the bodily source of the disease was not appreciated until 1870). A year after his recovery, Jenner felt ready to try the experiment he had been pondering so long. By chance an epidemic of cowpox appeared on the farm of a patient of his, and a milkmaid, Sarah Nelmes, became infected with true cowpox.

As well as details about Sarah (she was the daughter of a local landowner in nearby Breadstone), we also know about the cow. Her name was Blossom and she was a "Gloster" or Gloucester cow. Gloucester cattle are an ancient breed, numerous in the Severn Vale as early as the thirteenth century and valued for their meat and milk (producing a fine cheese) and as strong draft oxen. As a subject for his experiment, Jenner chose an eight-year-old boy who was in good health and had never had smallpox. His name was James Phipps, the son of Jenner's gardener. Jenner's intention was to take a small amount of the cowpox "virus" from Sarah's hand and place it under James's skin in order to try and protect him against smallpox.

Although Jenner used the term *virus* in his writings on cowpox, this was just another term for a poison or contagious principle. Viruses were not defined in any way until the early twentieth century. To Jenner and his generation, the infectious agents causing disease were shrouded in mystery, and this made it all the more important to distinguish carefully between the visible signs, symptoms, and all other objective characteristics of infection. Jenner was good at diagnosis, but as to the origins and relationships between infections, he didn't have much to go on. Jenner called cowpox the *Variolae Vaccinae,* or smallpox of cattle, by which he meant the cow equivalent of human smallpox. He believed that human smallpox and cowpox may both have been derived from "grease," an infection of horses—the human infection adapting over long periods during the domestication of animals, the cow infection by cross contamination when groomsmen also worked in dairies. Jenner wrote,

> There is a disease to which the Horse, from his state of domestication, is frequently subject. The Farriers have called it the Grease. It is an inflammation and swelling in the heel, from which issues "matter" possessing properties of a very peculiar kind, which seems capable of generating a disease in the human body . . . which bears so strong a resemblance to the small pox that I think it highly probable it may be the source of that disease.

This notion of a connection between cowpox and smallpox and their common source in an infection of horses provided some sort of rationale for Jenner's belief that true cowpox could protect against the dreaded human disease, but it was little more than speculation. A mild illness termed swinepox was also recognized at the time, and indeed the mild smallpox against which Jenner once sought to variolate his son and servants may have been confused with swinepox by the doctors in charge. These aspects only serve to emphasize the uncertainty surrounding the origin and identity of all infections at the time. The naming of diseases was based only on the appearance of symptoms, pocks being the marks of smallpox. And the name *Small Pox* was the term used to differentiate the infection from the Great Pox, syphilis, which produces a highly variable rash that is occasionally pustular like that of smallpox.

We now know that viruses *do* adapt to different species and that they *are* related to one another; but the only sound basis for understanding this lies in modern classification based on genetic sequence, mode of replication, host species, and disease pathology. Viruses evolve through changes in their DNA (or RNA); some evolve quite quickly, and the best-adapted mutants speedily outnumber their less fit counterparts. In this sense viral evolution is Darwinian, just like that of their hosts, but it occurs on a rapid timescale. Cowpox was originally a rodent virus, not a horse virus, and is a member of the same family of viruses as smallpox. Jenner had none of this information, but he was convinced that cowpox prevented smallpox, and he was determined to test his belief.

True cowpox infection takes the form of vesicles on the udder and teats of cows during lactation. These turn into pustules and then ulcerous localized sores, which are painful for the cow and make her difficult to milk. Transmitted fairly readily to milkers by skin contact through cuts and abrasions, it produces pustules on the hands and a mild malaise that soon gets better. Meticulous lifelike drawings of Sarah Nelmes' hand, bearing a few large, very distinct blisters, formed the centerpiece for Jenner's book when it was eventually published.

On May 14, 1796, Jenner was ready to proceed. He took some fluid (lymph) from a blister on Sarah's hand and transferred it on his lancet into the lower layers of James's skin (the same procedure as used for variolation). A painting depicts the fateful scene: Jenner sits upright, supporting James's extended arm, applying his needle to the upper surface. A woman in a mob cap (James's mother, perhaps) holds James firmly, soothingly cupping his chin while the anxious boy curls his fingers around her arm. Behind Jenner, Sarah stands watching with sympathy, holding a cloth to her hand where Jenner has just made his scratch. But this pleasant picture was imagined by the artist R. A. Thom in 1961, almost two centuries after the event. It appears in *Great Moments in Medicine* by George A. Bender (Park Davis, 1961).

For several days afterward, young James complained of discomfort around the armpit. Nine days after the inoculation he felt a little cold, lost his appetite, complained of a headache, and was indisposed for a day and a night. From then on he recovered and the inoculation spot faded. Both the lesion and the malaise were very mild compared to variolation. On July 1, some seven weeks later, Jenner variolated his small patient with smallpox material by the procedure he (and every other practitioner) routinely employed. He then watched anxiously for the result. To Jenner's great delight, James showed only a very small reaction of the kind seen in subjects already protected against smallpox.

Contrary to popular perception, Jenner did not take any unusual risks by exposing the boy to smallpox. As we've seen, exposure through variolation was the routine protective measure of the day. As to the risk from cowpox, the worst he could expect was a transient illness familiar to all in the countryside of Gloucestershire. On July 19, 1796, Jenner wrote to his great friend Edward Gardner,

I have at length accomlish'd what I have been so long waiting for, the passing of the vaccine Virus from one human being to another by the ordinary mode of inoculation.

A boy named James Phipps was inoculated on the arm from a pustule on the hand of a young woman who was infected by her Masters Cows. Having never seen the disease but in its casual way before, that is, when communicated from the Cow to the hand of the milker, I was astonish'd at the close resemblance of the Pustules in some of their stages to the variolous Pustules. But now listen to the most delightful part of my Story. The boy has since been inoculated for the smallpox which as I ventured to predict produce'd no effect.

The importance and originality of what Jenner did is best considered in three parts. Firstly, he realized that cowpox could deliberately be used to prevent smallpox infection. Secondly, he demonstrated that this could be done by transferring the infection from one person to another. Thirdly, and crucially, he confirmed by a well-established medical procedure that the measure had successfully protected against smallpox.

Jenner's discovery was a radical leap forward. It was already well-known that a mild episode of infectious illness protected against subsequent infection with the same disease; variolation was a contrived attempt to achieve this. While it was an important step forward, variolation produced highly variable results and the attendant danger of infecting others. But Jenner's realization, in rigorous intellectual terms—that one mysterious infectious entity (from a different species) could protect against another quite different infection—was an insight without precedent. In purely practical terms, the leap was even greater: the individual seeking protection no longer had to run the risk of serious disease, or even death, from the procedure. Nor did he have to risk the possibility of infecting others around him or transmitting the disease to those he loved.

In the first of the three aspects of Jenner's achievements there was an interesting precedent, though Jenner appears to have been unaware of it. In 1774 the pleasant village of Yetminster in Dorset was afflicted by a smallpox outbreak. A local farmer, Benjamin Jesty, feared for the safety of his wife and children. Jesty had already had cowpox, and knowing of the folk wisdom regarding the infection and convinced by his own experience, he resolved to protect his family. When an outbreak of cowpox took place in the neighboring village of Chetnole, Jesty equipped himself and took his family there. He searched out a suitable cow, took a darning needle and transferred pustular material from the cow's udder to the arms of his loved ones. The two boys had mild reactions and quickly recovered, but his wife, Elizabeth, was less fortunate. Her arm became swollen and highly inflamed, and for a time her condition gave cause for grave concern. (She probably had a secondary infection.) Happily, Elizabeth too eventually recovered.

Jesty's actions became generally known when he anxiously sought medical help for his wife, and as a result he suffered cruel persecution from his neighbors. At local markets he was hooted at and pelted and generally condemned for the disgusting act of transferring an animal disease into the human body. It took 31 years for Benjamin Jesty to be appreciated. In 1805, following representations from several learned men, Jesty was summoned to London to give evidence to the Vaccine Pock Institute. His eldest son went with him and agreed to be variolated to demonstrate that he was protected against smallpox. After receiving his account, the institute published Jesty's spoken evidence in the *Edinburgh Medical and Surgical Journal,* and he submitted to have his portrait painted.

But Edward Jenner's accomplishment was on quite another level. The three achievements of Jenner's first experiment—his insight into the protective potential of cowpox, his practical application by transferring the infection from one person to another, and his crucial demonstration that protection was successful—would mean little until a fourth achievement had been added. It was essential that, once he had collected enough information, Jenner should publish his work. It is a fact of science, then as now, that any discovery of nature's secrets barely exists until it has been shared with the scientific community with sufficient clarity and detail that others can repeat and independently confirm the findings.

5

THE FOURTH ACHIEVEMENT

"What can give us more sure knowledge than our senses? How else can we distinguish between the true and the false?"

Lucretius, First Century BC

These days it's called peer review: the rigorous and critical appraisal of scientific studies by experts in the field that constitutes the gateway to publication, guarding the quality and integrity of the scientific literature. It's very rare for any piece of scientific research to see the light of day until the review boards have had their say and seen their questions answered. Studies for publication are frequently rejected until the work has been further substantiated, often by lengthy additional experiments. The most widely read journals are also the most difficult to satisfy, so in order to have an impact it's essential to do work of high value and excellent quality. Waiting for those all-important editors' reports is a universal source of anxiety for modern scientists; the worst fear of all is the fear of rejection.

Jenner's crucial first experiment was in need of repetition and consolidation, but he faced an exasperating problem: no more cases of cowpox were reported in the Severn Vale. Instead much of his time was taken up in attending the young wife of Lord Berkeley who, for the first year after giving birth, suffered recurrent hemorrhages. Still, Jenner remained convinced and longed to tell the world of his discovery. In March and April of 1797 he wrote two versions of a manuscript describing his single experiment (together with the case histories he'd collected) and sent it off to the Royal Society.

He must have felt he had a good chance. After all, he was established as a man of science; his fellowship of the Royal Society guaranteed a fair hearing; and the enormous potential significance of his observations must surely ignite interest since the problem was so grave, especially in London where, despite variolation, the rates of smallpox remained ten times higher than in rural areas. Moreover, the man to whom he sent his paper was well known to him and already held him in esteem. That man was Sir Joseph Banks (the botanist on Captain Cook's great voyage whose collection Jenner had cataloged so successfully), now president of the Royal Society. All these things were in his favor as he waited to hear back.

In what might be considered a further piece of luck, when Banks had read the paper and thought about where to send it for a second opinion (such expert commentators are now called "referees") he settled on Sir Everard Home, Fellow of the Royal Society (FRS), who happened to be John Hunter's brother-in-law. Yet all this good fortune was to no avail: Home remained unconvinced. On April 22 he wrote to Banks, "If 20 or 30 children were inoculated for the Cow pox and afterwards for the Small pox without taking it, I might be led to change my opinion, at present however I want faith." Banks accordingly advised Jenner to withdraw his manuscript, suggesting more studies were needed to prove something "so much at variance with established knowledge."

Most experts would agree with Banks, and justice was probably done at this first attempt. Fortunately, in the spring of 1798, Lady Berkeley recovered and the weather turned wet in Berkeley Vale. Cowpox returned to the farms of Gloucestershire and Jenner resumed his studies. He went on to repeat his pioneering observation with several more cases, including cow to human vaccination and arm-to-arm vaccination from one child to the next, which he found just as efficient. He tested three of his ten fully documented vaccinees by subsequent variolation, confirming that protection against smallpox was effective (making four in all including James Phipps). He also confirmed the potency of the smallpox material used in the variolations by testing it in an unvaccinated person. Now he was ready to produce a more mature report. But, perhaps fearing another rejection, Jenner ignored the Royal Society and decided to publish privately, and at his own expense, *An Inquiry into the Causes and Effect of the Variolae Vaccinae—A Disease Discovered in Some of the Western Counties of England, Particularly Gloucestershire, and Known by the Name of the Cow Pox.*

Unlike its title, the book was concise—75 pages in large print with four lavish colored engravings—dedicated to his friend Caleb Parry, priced at seven shillings and sixpence (around £40 in today's money), and was sold at two London bookshops. The text begins poetically, speaking of man's fondness for domestic animals with portraits of our two favorites: dogs and cats. "The Wolf disarmed of ferocity, is now pillowed in the lady's lap. The Cat, the little tyger of our island, whose natural home is the forest, is equally domesticated and caressed." He then moves swiftly on to cows. In terms of sales, the book, published on September 17, 1798, was destined to become a huge success. As for the term *vaccination,* it does not appear in the book and was never used at this time. It would be some years before Richard Dunning, a Plymouth surgeon and friend of Jenner, coined the term (from *Vacca,* the Latin for cow) in Jenner's honor.

When Jenner took his manuscript to the publisher (Samson Low in Soho), he also took with him to London a supply of vaccine from the arm of a patient, dried and sealed in the quill of a feather. He anxiously hoped to demonstrate its effectiveness to the London medical establishment. But he was to be disappointed. His celebrated mentor, John Hunter, had been dead five years, and no one showed the least bit of interest. After three months Jenner returned despondently to Berkeley, leaving the material in the care of his old friend Henry Cline.

Two weeks later he received an enthusiastic letter. Cline had, in a sense, misused the precious vaccine by administering it to a boy with an inflamed hip, hoping a counterirritation might help the child's condition. It didn't, but fortunately Cline then decided he might as well see if the vaccine had worked in the way Jenner said it would. He variolated the boy and found he was protected against smallpox infection. At once he wrote to Jenner:

> The cowpox experiment had succeeded admirably . . . Dr. Lister, who was formerly physician to the smallpox hospital, attended the child with me and he is convinced that it is not possible to give him the small-pox. I think the substituting of the cow-pox for the smallpox promises to be one of the greatest improvements that has ever been made in medicine; and the more I think on the subject, the more I am impressed with its importance.

For Jenner, back at Chantry Cottage, prospects for the future suddenly looked bright.

Cline's enthusiasm for cowpox inoculation caused quite a stir. Despite variolation, smallpox still killed 2,000 people every year in London. Plenty of publicity followed, and things began to move at last in the London medical establishment. In particular, George Pearson, physician at St. George's Hospital, began his own investigations, and in November he published *An inquiry concerning the history of the cow pox. Principally with a view to supersede and extinguish the smallpox.* It was exactly what Jenner needed, given that Pearson was careful to disclaim originality and fully acknowledge the inspiration for his work: "Perhaps it may be right to declare, that I entertain not the most distant expectation of participating in the smallest share of honour, on the score of discovery of facts. The honour on this account, by the justest title, belongs exclusively to Dr. Jenner; and I would not pluck a sprig of laurel from the wreath that decorates his brow."

Pearson's study was a general collection of case histories where immunity to smallpox followed cowpox infection, together with his own experience of five patients who'd had cowpox and were resistant to smallpox infection (by variolation). Quite soon afterward outbreaks of cowpox in a dairy near Gray's Inn and in a herd at Marylebone Fields meant Pearson could begin his own experiments. This he did with William Woodville of the London Smallpox Hospital.

Although he wrote to Jenner and sent him vaccine material, Pearson's energetic activities in the heart of London and his network of medical communications soon threatened to overshadow Jenner's in the quiet vales of Gloucestershire. In March 1799 Jenner's nephew George warned his uncle that Pearson's reputation, as the sole supplier of cowpox material to London medical men, might soon eclipse his own. In response, Jenner came once more to London, published his follow-up studies in *Further Observations on the Variolae Vaccinae or Cow Pox,* and stayed for three months. But trouble of the worst kind lay ahead.

The problem was that Pearson and Woodville were performing their studies of vaccination at the London Smallpox Hospital. Since it was vital that cowpox material remain pure and free from contamination, especially from smallpox itself, the hospital was perhaps the last place on earth in which to mount the study. Of course, the standards of the day were in place to prevent contamination, but they simply weren't enough. Of 500 cowpox vaccinations, more than half resulted in the numerous pustules characteristic of smallpox variolation. Worse, these multiple pustules were sometimes used as the source for more vaccine. The problem was compounded by Woodville and Pearson's roles as vaccine suppliers to London (and to Europe and North America); they sent out vaccine dried onto threads, some of which came from the London Smallpox Hospital. Variolation was still very much in use at this time—about half the people seeking protection still chose the traditional method, and cowpox and smallpox material was sometimes prepared on the same table. No wonder the results of vaccination seemed so variable! The subsequent fortunes of disease prevention would have many such episodes of haste leading to a lack of care, perhaps the most tragic being the contamination of polio vaccine in its first field trials 150 years later.

Despairing of the London scene, Jenner returned to his beloved Gloucestershire, resolved to champion the selection of only true cowpox as a vaccine material. He was determined both to vaccinate the local poor and to supply pure, authentic cowpox throughout the land. But then things turned from bad to worse.

From the start Pearson, who still strongly believed in vaccination despite the limitations his studies suggested, set up the Institution for the Inoculation of the Vaccine Pock (the term *vaccination* was not yet in use) as a national center for cowpox inoculation and had the gall to offer Jenner the all but meaningless post of "extra consulting physician."

It is perhaps rather easy to characterize Pearson as a usurper, and an incompetent one at that. The fact is, he did appreciate the value of vaccination; the quality of London vaccine was eventually improved, and he did support and champion Jenner's aims. Other figures were much more troubling for Jenner, and for vaccination.

In 1828 a Newcastle surgeon, Henry Edmonston, wrote of how Jenner's discovery was received with joy and gladness in some quarters and raillery and abuse in others, echoing public reactions to variolation some 80 years before Jenner's breakthrough. Jenner's detractors were many and varied and included Dr. Benjamin Mosley, who wrote, "Can any person say what may be the consequences of introducing the *Lues Bovilla* [syphilis of oxen], a bestial humour—into the human frame . . . ? Who knows, also, that the human character may undergo strange mutations from quadruped sympathy . . . ?" He went on to suggest that vaccinated ladies might find themselves wandering in the fields to receive the embraces of the bull. Dr. Thomas Rowley claimed that vaccination had produced an "ox-faced" baby boy, while John Birch of St. George's Hospital denounced it as useless and predicted that replacing variolation with vaccination would allow smallpox to "recur in all the terrors with which it was first surrounded." The frenzied fear of vaccination as a form of ungodly bestial degeneration prompted the political cartoonist James Gillray to mock the hysteria in a famous cartoon of 1802. Fashionable folk are shown aghast at the ox-headed monsters violently erupting through their skin as a result of vaccination.

The vehemence and extreme beliefs of those opposing vaccination were, of course, never justified, but there were more reasonable critics yet to be convinced. Part of the problem can be traced to Jenner's self-published *Inquiry*, which bypassed the improvement it might have received at the hands of expert reviewers. One of the difficulties was Jenner's assertion that cowpox originated as the infection of horses called "grease," a speculation without evidence that detracted from the much more important practical aspects. Another was the sample size, still quite small, of ten named cases of cowpox inoculation (including a short chain of six human-to-human transfers), only four of which were variolated (though all four were protected). A third weakness was Jenner's inconsistency about the size of the cowpox lesions at vaccination sites, whose severity he sometime sought to minimize. Despite the crucial importance of identifying true cowpox for vaccination material, he did not include illustrations of bovine pustules (though he described them in detail). He was evasive about one unfortunate subject, John Baker, whom he said could not be variolated after vaccination "from having felt the effects of a contagious fever in a workhouse." In his subsequent publication *Further Observations,* Jenner reveals in a footnote that Baker died. While this was probably due to unrelated causes, Jenner was unwilling to discuss the subject. Also, Jenner stood firm on his assertion that cowpox infection and vaccination gave lifelong protection against smallpox. All cases of infection post vaccination, he claimed, could only be due to the use of spurious cowpox. In the years that followed, it gradually emerged that Jenner was wrong on this point and that protection waned with time (as it does with all vaccines). Having to defend this mistaken viewpoint substantially weakened the otherwise excellent case for vaccination.

To minimize the negative features, Jenner played up the positive aspects of his case. He was also obstinate and overly sensitive to criticism. The great strengths that Jenner brought to his study included his quite exceptional powers of observation and his insights regarding the nature of infection, which were well in advance of his time. Without knowing anything about viruses or germs, he was able to visualize their characteristics, grasping that different infectious agents might produce similar symptoms, that different agents might occur in the same lesions in a consistent sequence, and that the agents of infection might become weakened or lost in tissues over time and in storage.

In relation to vaccination, and to Jenner, four camps emerged. Of those in favor, there were supporters of both Jenner and vaccination, and there were supporters of vaccination who sought to minimize Jenner's contribution. Among the latter were would-be claimants and usurpers. Of those against, there were those not convinced by the evidence, some of whom had vested interests in variolation, and there were passionate enemies, many of them fervently opposed on philosophical, religious, or moral grounds. While Jenner complained to Edward Gardner of the "Brickbats and hostile weapons of every sort that are flying thick around me," he asserted that "there will be no end to cavil and controversy until it be defined with precision what is, and what is not Cow Pox."

To this end, in *Further Observations* (1799), Jenner warned against spurious cowpox, including the bovine pustules of pseudocowpox,

horse-derived material that lacked potency, and matter from older pustules, where the true virus was gone and other infections had taken over.
However, horsepox was later sometimes used, alongside cowpox, as a
source of vaccine. He also warned of loss of vaccine potency in storage.
Only when true cowpox from fresh pustules was used would vaccination
succeed.

In addition to producing only a minor skin lesion, which greatly
reduced fever and malaise and healed in a shorter time, cowpox vaccination had another great advantage over variolation: it did away
completely with the risk of smallpox spreading to others during the inoculation period. This was a pivotal public benefit of the new practice,
and Jenner's supporters were soon enthusiastically campaigning for its
adoption.

One of them was Dr. John Coakley Lettsome, the influential Quaker
doctor, reformer, and philanthropist. He had championed variolation
and supported the Society for the Inoculation of the Poor but was soon
converted to the advantages of vaccination. He sent a copy of Jenner's
Inquiry to fellow Quaker Benjamin Waterhouse at Harvard University,
and Waterhouse became the first major practitioner of vaccination in
America.

Jenner's friend John Ring, of the Medical Society of London
(founded by Lettsome), also became a tireless proponent. From 1799,
Ring devoted himself to the cause of vaccination, investigating every
adverse case that he heard of in London, rebutting Jenner's critics, of-
fering free vaccination to all who wanted it, and organizing a petition in
support of vaccination across the London establishment. The evangelist
preacher Rowland Hill also took up the cause, publishing a promotional
pamphlet and vaccinating first hundreds and eventually thousands of
people who came to hear him preach. This prompted Lettsome to write
in praise of him, "You have done more good than you imagine; and for
everyone you may have saved by your actual operation, you have saved
ten by your example; and perhaps, next to Jenner, have been the means
of saving more lives than any other individual."

Beyond London, the practice was also spreading. An outbreak of
smallpox in Westmorland in 1800 was halted by the local doctor, Robert
Thornton, who vaccinated all 400 villagers and then a thousand more in
the vicinity. In 1802, a Mr. Thomas of Daventry wrote to a learned journal and wryly observed, "To my very great disappointment I found the
Vaccine- pock so safe and mild a disease . . . that I became a convert."
He went on to say he was compelled to close his inoculating houses.

As the virtues of vaccination became established through the efforts of individual supporters, Jenner began to receive recognition for

his achievement, though he still had plenty of opponents. In May 1800 he received permission to dedicate the second edition of the *Inquiry* to King George, and he was twice invited to St. James Palace to meet the king and queen. He was also invited to Oxford University, where the medical faculty presented Jenner with a testimonial: "We, whose names are undersigned, are fully satisfied upon the conviction of our own observation, that the Cow pox is not only an infinitely milder disease than the Small Pox, but has the advantage of not being contagious, and is an effectual remedy against the Small Pox."

Two years later, urged on by his supporters, Jenner petitioned Parliament for some remuneration for what he had achieved, pointing out the great benefits the country had enjoyed and the fact that he had shared his findings and not sought profit or payment for his discovery. It was a normal practice of government to reward significant achievements in this way. John Harrison, for example, was awarded nearly £8,000 in 1773 for his extraordinary work on marine chronometers for finding longitude at sea. We should remember too that variolaters made substantial fortunes, not least Thomas Dimsdale, who received £12,000 from Catherine the Great and a pension for life. The king warmly supported Jenner, and Lord Berkeley was appointed chairman of a committee of inquiry.

What kind of reward might be appropriate to the problem whose solution seemed at hand: a scourge that damaged empires, decimated nations, and ruined economies? Lord Sherbourne of Gloucestershire felt Jenner ought to have at least £50,000. Others thought £20,000 would be appropriate. (Jenner would have made a larger sum by seeking profit from his vaccine.) Many witnesses were called, and all attested to the originality and enormous value of Jenner's work and his great generosity and lack of self-interest. Everything was going extremely well. And then George Pearson was summoned to give evidence.

Pearson told a very different story. He did his utmost to belittle Jenner's contribution, citing the exploits of a certain farmer named Jesty of Dorsetshire 22 years before Jenner's experiment. In fact it was Pearson, motivated by a newfound antipathy toward Jenner, who belatedly arranged for Benjamin Jesty to be recognized by the Vaccine Pock Institute in 1805. Pearson also claimed that he and Woodville, not Jenner, had proven the effectiveness of vaccination. (Woodville declined to be associated with this claim.) The tenor of the debate changed, and lengthy discussions ensued. At last the decision was reached and justice prevailed. An agreement to award Jenner £10,000 was secured by a majority vote. It was decided that he was the "sole discoverer" of what was judged to be "the greatest discovery ever made in medicine."

The importance of these honors and awards went far beyond personal vindication for Jenner. Acceptance of the value of his work at the highest levels of British society provided the foundations on which the vaccine movement would extend its influence, in the United Kingdom and far across the globe.

6

THE FOUNDLING
VOYAGES

*"No man has suffered more . . . from the excessive eulogies of his
friends or from the invective of those who did not subscribe to his
views."*

E. A. Underwood, 1949

On March 2, 1803, a society for the extermination of smallpox
through vaccination was inaugurated in London. It came to
be called the Royal Jennerian Institution, and it distributed
tens of thousands of doses of vaccine for five years until it was superseded in 1808 by the National Vaccine Institute. Jenner, determined to
secure the advance of vaccination practice, unstintingly used the money
granted him to fund the work of the institution in London and for other
public initiatives; his personal budget remained very small. Much of his
time was taken up dealing with a growing worldwide correspondence
and advising vaccinators around the globe. In August 1803, while honors rained down on him, he found himself, as we might put it, somewhat
strapped for cash. "Honours pour upon me," he wrote, "but honours
do not buy mutton." He returned to Gloucestershire where he could
conduct his correspondence from the comforts of home.

More honors were to follow. In July 1805 Jenner received the Freedom of the City of London, which came in a gold box worth 100 guineas. In response he said the privilege made him extremely happy, not
least because the testimony of the Court of Common Council would
counteract attempts, originating out of "ignorance or prejudice," to re-
tard the progress of vaccination.

A year later the House of Commons once more considered Jenner's case. Severe outbreaks of smallpox were focusing attention on the importance of his achievements. The king asked the Royal College of Surgeons and the Royal College of Physicians to provide their opinions, and the latter, after a thorough enquiry, produced a report strongly and without reservation recommending Jennerian vaccination. Parliament debated and, by 60 votes to 47, decided that Jenner would receive a personal award of £20,000. (This was not a huge sum at a time when English country gentlemen such as Jane Austin's Mr. Darcy "had £10,000 a year.")

When the National Vaccine Institute was created in 1808, Jenner was appointed director. Isaac Cruikshank's contemporary cartoon shows Jenner accompanied by two colleagues holding his vaccination scalpel inscribed with the words "milk of human kindness." Retreating from his advance are three grotesque variolaters grasping their giant instruments dripping with purulent material. One says, "Curse on these vaccinators. We shall all be starved, why Brother I have enough here to kill 50." The second adds, "And these would communicate it to 500 more." "Aye Aye," the third avers, "I always order them to be constantly out in the air, in order to spread the contagion." Around their feet lie the bodies of dead infants (perhaps it is essential to demonize a once useful practice as soon as it is superseded). Although Jenner was made director, there was still a crucial problem: because he was not a fellow of the Royal College of Physicians, he could not be a member of the administrative council that ran the institute. His influence on how vaccination campaigns were conducted was limited, and he resigned in frustration.

It would be five years before Jenner was officially honored with the degree of Doctor of Medicine from Oxford University, but even that wasn't good enough for the Royal College of Physicians. Because Jenner still lacked credentials in the classical languages, there was no way he could be admitted as a fellow. "In my youth," Jenner said (according to biographer John Baron), "I went through the ordinary course of classical education . . . but the greater part of it has long since transmigrated into heads better suited for its cultivation. At my time of life to set about brushing up would be irksome to me beyond measure: I would not do it for a diadem."

In the year the *Inquiry* was published, the French Revolution began, but the news of Jenner's book reached France despite the turmoil. The Parisian School of Medicine began an investigation into the merits of the new treatment, and French practitioners were soon receiving vaccine from Pearson and Woodville. By far the biggest problem in trying to spread vaccination through Europe and beyond was not the almost constant state of war; it was the problems of vaccine instability, which

Jenner had always recognized. Vaccination was tested out on a group of orphans at the Hôpital de la Pitié in Paris using material sent by Pearson, but it failed to induce a proper reaction. It had almost certainly decayed in the hot weather. The French then issued a special passport for William Woodville so he could bring fresh vaccine from London. When Woodville arrived in June 1800, he was accompanied by Dr. Thomas Newell, an Englishman who practiced in France. By the time they reached Paris, the vaccine had again lost potency. Luckily Newell had successfully vaccinated three children in Boulogne on his way from London to Paris, and he returned to fetch fresh material from these small human depots. Gradually they began to spread the practice in France and neighboring countries. Their efforts led to the formation of the Comite Central de Vaccine in Paris, which subsequently oversaw the organized establishment of vaccination at the national level.

At the time of the Comite's establishment, there were three consuls governing France, and one of them was quick to realize the value of vaccination. His name was Napoleon Bonaparte. When Napoleon became emperor he ordered that 100,000 francs should be spent to encourage and propagate vaccination. He was quick to follow the British policy, and in 1805 arm-to-arm vaccination was introduced for the troops of the emperor, although this wasn't rigidly enforced until 1811. In true Gallic style he had a medal struck featuring his image with Aesculapius (Greek god of medicine) and Venus on the reverse, flanked by a small cow and scalpel. The cow looks remarkably like Blossom, the original donor of cowpox vaccine. In the year of the Battle of Waterloo (1815), the subject of the French Academy's poetry competition was "Edward Jenner and Vaccination."

Early in 1805 Jenner addressed himself directly to Napoleon, humbly requesting that two of his friends, Mr. Williams and Dr. Wickham, both men of science and literature who were imprisoned when the war broke out, might be allowed to return to England. According to Jenner's friend Fosbroke, this happened at the time of Napoleon's greatest animosity toward Britain. "The Emperor was in his carriage," Fosbroke wrote, "and the horses were being changed. The petition was then presented to him. He exclaimed 'Away, Away!' The Empress Josephine who accompanied him said 'But Emperor, do you see who this comes from? *Jenner.*' He [Napoleon] changed his tone of voice that instant and said 'What that man asks is not to be refused,' and the petition was immediately granted."

The fame and honor accorded the practice of vaccination in Napoleon's empire reached its height at Chateau de Saint Cloud on May 11, 1811, with the vaccination of the King of Rome, Napoleon's two-month-old son upon whose infant head the hopes of the dynasty rested.

"The whole Empire knows today that His Majesty, the Emperor and King, adopted for his august son, the King of Rome, the salutary method of the vaccine." Two days later Jenner was elected a foreign associate member of the Institute of France; the news was communicated to him by Sir Joseph Banks in his role as president of the Royal Society.

Elsewhere, from early in the century, the principle of arm-to-arm vaccination was used to convey effective vaccine across the world. Dr. Jean de Carro, working in Vienna, vaccinated his own children and confirmed success by challenging them with variolation. He then became a champion for the spread of vaccination across the globe. Dr. Luigi Sacco, an influential Italian convert to vaccination, discovered an outbreak of cowpox in cattle in Lombardy, and using this material, Carro initiated an arm-to-arm human chain of fresh vaccine across Greece and Turkey to India. Encouraged by Lord Elgin, the British ambassador to the Ottoman Empire, the chain proceeded via Constantinople where, 80 years earlier, Lady Mary Wortley Montagu had first encountered variolation.

From Baghdad this vital human procession took on the form of vaccinated crewmen and passengers on a ship bound for Basra. From there the human chain continued to Bombay where a three-year-old girl, Ann Dusthall, was the first to be vaccinated on Indian soil in 1802. She alone was then to become the primary source for vaccine material snaking chain-wise throughout the Indian subcontinent. Within three years, 30,000 Ceylonese had been vaccinated, and in 1821 endemic smallpox on the island of Ceylon had all but disappeared.

An even more carefully orchestrated arm-to-arm chain was organized through the enlightened enthusiasm of King Charles IV of Spain. In 1803 he commissioned the Royal physician, Francisco Xavier Balmis, to undertake a voyage fully equipped with medical staff together with 22 foundling children ages three to nine. One of the orphans was vaccinated just before the ship set sail. Thereafter, at ten-day intervals, two more children were vaccinated with fresh pustular material. Balmis first visited the Canary Islands, establishing a vaccination clinic there before sailing on across the Atlantic. When at last he reached Caracas in Venezuela, only one boy still had fresh pustules. But one boy was enough! Vaccination was launched on the South American continent nearly three centuries after the Spanish invaders had carried smallpox to the New World.

But what happened to the foundlings? Happily they were settled in Mexico, educated at the expense of the Spanish treasury, and eventually adopted by local families. Balmis's humane and unprecedented mission did not end there. The expedition was divided into three, and the vaccine net was cast as wide as Peru, Buenos Aries, Chile, and the Philippines. A crucial part of Balmis's plan was to set up vaccination boards to ensure that vaccination flourished, and, with keen support from church and civil authorities, his mission succeeded. Balmis himself led the expedition to Cuba, Yucatan, and Mexico. The destinations for his human chain eventually included Manila, Macao, Canton, and at last St. Helena on the long voyage home. The mission to the Philippines was also a long one: a 26-boy voyage was needed to maintain fresh material for such a great journey. The Mexican boys' reward was the promise of a free education when at last the long trick was over. In Manila, the calves of the native water buffalo, *caraboa,* were used to produce more vaccine, which was freely available to citizens and shipped out to the provinces in glycerin sealed in tubes and bottles. Balmis's great journey lasted almost three years. Jenner called it a "glorious interprise" and told his friend Richard Phillips, "I have made peace with Spain and quite adore her philanthropic monarch." (Britain was at war with Spain at the time.)

Balmis's achievement was unprecedented, but it was not to be unique. The Portuguese were quick to adopt the Spanish model for their own prize colony Bahia (in Brazil). In 1804 slave boys from Bahia were transported to Lisbon and then used to form an arm-to-arm vaccine chain for the long voyage back to the New World. In the annals of sea-faring, these great voyages must surely be among the strangest and most significant ever undertaken. The boys' fates seem to be unrecorded, and though they were returning whence they came, their real home—the homes of their families—lay in Africa.

Remarkably, the almost constant state of war in some ways helped the march of vaccination. In 1800, Frederick, Duke of York, sent two physicians, Joseph Marshall and John Walker, to organize the protection of troops against smallpox in the Mediterranean theater. The two men carried a chain of vaccinated sailors through Gibraltar and Minorca to Malta. Walker then continued into Egypt, where British troops fighting the French were vaccinated. Marshall sailed on to Sicily and set up a vaccination center in Palermo before traveling to Naples, where he established an Institution for Jennerian Vaccination. In the rare years of uneasy peace in 1802 and 1803, Marshall seized the opportunity to journey all the way to Paris, passing through Rome, Genoa, and Turin, planting vaccination clinics and training vaccinators as he went.

In northern Italy the tireless vaccinator Luigi Sacco had already been at work. He had become director of vaccination for Napoleon's Cisalpine Republic in 1801, and he eventually succeeded in putting an end to endemic smallpox in Lombardy, personally vaccinating many thousands and organizing vaccination for many, many more. To reach the more remote Alpine regions, a newly vaccinated boy traveled with the practitioner as a vital source of fresh vaccine. Sacco proudly claimed that his region of Italy had become preeminent among European kingdoms in promoting vaccination. When smallpox broke out in Rome in 1814,

Pope Pius VII endorsed the vaccination program; by 1824 Sacco was able to tell John Baron that almost all newborn children in the state of Lombardy received vaccination.

The records of smallpox deaths tell the story of the success of vaccination. From 1812 to 1821, total smallpox deaths in Berlin fell from 5,000 to less than 600, assisted by the incentives of the Royal Inoculation Institute, which gave out medals to poor children. The king of Prussia, Friedrich Wilhelm III, set a fine example in 1799 when he had his children vaccinated just one year after Jenner's publication. In Sweden the records show that deaths from smallpox had fallen more than tenfold by 1821.

In the wider world, progress continued more spasmodically, with disastrous epidemics in the colonies continuing to decimate the poor and indigenous peoples of the world. In Cape Town around 2,500 died in 1840, and there was little vaccination in sub-Saharan Africa, although slave traders soon learned the value of vaccination for their Atlantic trade. It wasn't until the twentieth century that vaccination began to spread across Africa.

The story in North America began very soon after Jenner's *Inquiry* was published. A childhood friend from Jenner's days in Cirencester, John Clinch, had become a missionary doctor in Newfoundland. Jenner sent him a packet of dried vaccine in 1798. Clinch evidently trusted his old school friend, for he quickly vaccinated his nephew and arranged for him to share a bed with one of his own highly infectious smallpox patients. To everyone's amazement the nephew remained well. Soon everybody was demanding the miraculous treatment. Clinch began a chain of vaccinations—the first in North America—that reached 700 until the chain was unfortunately broken. New supplies arrived in 1802, but it was not until 1821 that a Bureau de Vaccine was established in Quebec.

Further south, in Boston, the physician Benjamin Waterhouse had received a copy of the *Inquiry* from Jenner's advocate John Lettsome in London. Keen to investigate, Waterhouse obtained material from the English physician and researcher John Haygarth. Waterhouse vaccinated his son and six household members, then tested for protection by variolation with smallpox. When he saw the effectiveness of the procedure, he conceived a plan to supply vaccine to a network of practitioners, selling the material for his own gain and maintaining a monopoly to maximize his income. The result of this pursuit of profit led to a black market in vaccine material with little control of quality. Disaster soon followed: a sailor who was used as a source of cowpox vaccine turned out to be suffering from smallpox. More than 60 vaccinated people died.

Following the outcry, Waterhouse relented and began to freely distribute authentic vaccine, promoting the practice in a public-spirited way. It was Waterhouse who persuaded the Boston Board of Health to vaccinate nineteen volunteers and test for protection by variolation, hoping that the publicity surrounding an official demonstration would encourage the new treatment. His plan succeeded: small communities took up the practice, laws were passed encouraging vaccination, and between 1816 and 1824 there were no deaths from smallpox in Boston.

Soon after the success of his initial investigations, Waterhouse wrote to Thomas Jefferson, who had just been elected as president, about the possibilities of vaccination. Jefferson received the letter on Christmas Eve 1800, read it that night, and wrote to Waterhouse on Christmas Day: "Every friend of humanity must look with pleasure on this discovery, by which one evil more is withdrawn from the condition of man; and must contemplate the possibility, that future improvements and discoveries may still more and more lessen the catalogue of evils."

When Jefferson first received vaccine material from Waterhouse in the hot summer of 1801, he arranged for it to be tested urgently. It turned out to be ineffective, producing no pustules. The same thing happened with two more batches dispatched by Waterhouse. Far from being discouraged, Jefferson designed an insulated package in which vials could be immersed in water to keep them cool. In August 1801 he wrote from Monticello, his home in Virginia, that a number of inoculations, including twenty in his own family, had succeeded. At first a local physician, Dr. Wardlaw, carried out the vaccinations, but when he was too busy the president took over. Waterhouse would soon advise that there was no need to send more vaccine because the campaign could continue with the Virginia material. In August and September Jefferson sent vaccine to Petersburg, Richmond, and other parts of the state where earlier failed vaccinations had left people skeptical about the procedure. By the end of 1801 he had introduced vaccination across Virginia as well as in Philadelphia and Washington, D.C. He even went as far as to ensure that a proportion of the vaccinations he carried out were proved to be effective by subsequent variolation. These achievements would be admirable for any scientific investigator, but for a president of the United States they were, and surely must remain, unique. He went on to promote the vaccination of Native American people across the continent through the expedition of Lewis and Clark and other initiatives; local tribes were told that the enlightenment of vaccination came to man from the "Great Spirit." On May 14, 1806, during his second term as president, Jefferson wrote to Jenner, "You have erased from the calendar of human afflictions one of its greatest. Yours is the comfortable reflection that mankind can never forget that you have lived. Future nations will know by history only that the loathsome smallpox has existed and by you has been extirpated."

7

THE TEEMING
HUMANITY OF
NATIONS

*"Moreover, these flexions are taking place every where, like a simul-
taneous motion of all the waves of water of the world: and these
are the classic patterns, and this is the weaving, of human living: of
whose fabric each individual is a part."*

James Agee, *Let Us Now Praise Famous Men*, 1941

From the outset, Jenner himself had believed in the possibility of
global eradication, writing in 1801,

A hundred thousand persons, upon the smallest computation,
have been inoculated in these realms. The number who have partaken
of its benefits throughout Europe and other parts of the Globe are in-
calculable; and it now becomes too manifest to admit of contradiction
that the annihilation of the smallpox, the most dreadful scourge of the
human species, must be the final result of this practice.

Yet the gulf between the ideal and the reality was unimaginably vast.
Three things were on Jenner's side: smallpox existed nowhere but in
humans, victims were infectious for only a short time, and those who
recovered were immune for the rest of their lives. Against him was the
fact that, even in isolated civilizations, wherever there are cities smallpox
can survive forever, as it did in the tenth century walled city of Baghdad,
where Al Rhazi described it repeatedly afflicting each vulnerable new
generation.

There are many ways to portray the spread of vaccination across the globe, and the simplest and most inspiring is to focus on success: the glorious enterprise of vaccine voyages, the dedication of pioneers, the determination of intrepid travelers paying out the chains of vaccination arm to arm until a net was cast across the globe. But in another view, the annihilation of smallpox looks impossible. No region of the populated world must be left untreated, and the global vaccine net must be turned into something much more like a blanket. The problems faced by any such campaign are as manifold and diverse as the geography of the earth and the teeming humanity of all its nations.

Carrying vaccination into the interior of the most populous countries was exceedingly difficult. There were kingdoms where the great majority of people were illiterate, poor, ethnically and culturally unrelated, spread out across enormous territories, speaking different languages. Some of the greatest problems had to do not with distance and logistics but with beliefs. And, most difficult of all, certain countries were closed to outside influence.

In India, where British rule facilitated vaccination programs, smallpox was an integral element of religious beliefs. The Hindu goddess of smallpox, Sitala or Shitala Mata, was worshipped by Hindu, Buddhist, and tribal peoples throughout the subcontinent. Sitala is usually depicted as a mild and comely woman sitting sidesaddle on a donkey, carrying a broom and a pitcher of water. Her name means "the cool one." Sitala is crowned with a winnowing fan, and though she scatters the seeds of smallpox, she also brings relief with cooling water and a cleansing brush. In her other guise she is a fearsome goddess with daggers raised above her head.

Vaccination would seem to interfere in the divine relationship between the goddess Sitala and her people, but many ingenious strategies of persuasion were used to make the new practice acceptable in India. A Sanskrit poem on vaccination was contrived in the semblance of an ancient teaching to help persuade the reluctant. Richard Wellesley, the governor general, provided extra payments to those traditional Brahmin variolaters who switched to vaccination. The East India Company was also active in encouraging vaccination because it was extremely good for trade. And the new blessing came too from a sacred animal, the cow, though this may also have been a difficulty given the revered status of the vaccine donor. The Princely States of India likewise began to adopt their own independent vaccination measures.

Today, with smallpox banished from the world, Shitala Mata's temples remain important—she is the goddess of skin diseases, but she carries with her important remedies, one of which is the sacred tree, *neem*, a source of medicines for skin disorders in Ayurvedic medicine and a source for antifungal and pest-control chemicals.

In China, the ancient home of variolation, Jennerian vaccination arrived by several routes, one of which was Balmis's great mission to Canton in the south. Another was through Portuguese practitioners at Macao west of the Pearl River Delta (the first and also the last European colony in China). Alexander Pearson, surgeon to the East Indian Company in the early nineteenth century, practiced vaccination in Canton Province. At first, as Jenner would have wanted, he treated the poor at his own expense, but when the benefits became evident Chinese conservatism transformed into Chinese zeal, spurred on by the smallpox epidemic of 1805–6. When the threat of smallpox receded, Pearson had difficulty maintaining arm-to-arm supplies of vaccine, but in 1815 a free dispensary, run by the Chinese, was set up in Canton. When Jenner received a Chinese pamphlet on vaccination in 1806, he wrote to Richard Dunning, "Little did I think, my friend, that Heaven had in store for me such abundant happiness." China is a good example of the long ribbon of vaccination quickly cast around the globe and the enormously slow and difficult establishment of blanket vaccination. According to a retrospective account in the *British Medical Journal* of 1907, pock-marked people were the rule north of the Yangtze River until late in the nineteenth century. Parents were unwilling for their children to marry anyone who hadn't had smallpox. Variolation was still widely practiced even into the twentieth century, and when vaccine was in short supply variolation was substituted in rural regions right up to 1965. This was also true in Africa, Afghanistan, and parts of Pakistan. Variolation was made illegal in British India but was still practiced in some of the princely states. In the most remote kingdoms, such as Tibet, vaccination did not arrive until the 1940s.

At the time of Jenner's discovery, Japan was a mysterious nation closed to the West, although news of vaccination reached Nagasaki in 1803. In 1820 a book on Jennerian vaccination was translated from Russian into Japanese, and in 1848 a Dutch doctor, Otto Mohniké, arrived in Nagasaki with vaccine material. Despite such a late start, vaccination was quickly taken up in Japan. In 1858 a central service was established in Tokyo, and this organization became an important base for European treatments as opposed to traditional medicine. Initially private, it became a public Institution, and in 1861 it was transformed into a research establishment focused on Western medicine. Eventually it was directed by the eminent Japanese physician and bacteriologist Kitasato Shibasaburō, and it became the current Faculty of Medicine of the University of Tokyo. The Mikado and his wife supported the new practice and were vaccinated in 1875, in time to escape an epidemic of smallpox that killed the eighteen-year-old emperor of China. In 1879 a Japanese mission was sent to Europe to study the production of vaccine, and in 1885 vaccination was made obligatory.

Despite the high ideal of worldwide eradication that Jenner and his influential supporters conceived early on, there were many serious impediments to the much more modest aim of controlling smallpox back at home. One major difficulty was that variolation continued alongside vaccination, frustrating the effectiveness of the new practice. Despite several attempts at legislation, the outdated procedure of variolation was not made illegal until 1840.

Events in Norwich in 1818 tragically illustrate the problem. The town had been free of smallpox for five years, and less than a fifth of the population had been vaccinated. A family arrived from York, bringing with them their sick daughter. The little girl had smallpox, and she infected two more people. In response a pharmacist variolated three children, and the contagion spread from these individuals until more than 3,000 were infected. The outbreak killed 530 people.

But in the neighboring borough of Thetford, the authorities were determined to halt the spread of disease. John Cross, who wrote a history of the Norfolk "Variolous Epidemic" in 1820, recorded that the parish officers visited every house, listing those susceptible (not vaccinated) and threatening to expose anyone who refused vaccination or who sought clandestine variolation. Those unprotected were summoned by church bells to receive vaccination in their parish. In the course of two days, around 200 people were vaccinated, and less than a dozen people became infected. All of them survived.

In 1818 smallpox broke out in Edinburgh, and here the problem for the vaccination movement was much more serious—1,500 of the smallpox cases were purported to be in vaccinated individuals, three of whom died. Confidence in the new measure was in jeopardy. Jenner blamed the trouble on incompetent vaccinators, and J. F. Marston, surgeon at the London Smallpox Hospital, agreed. Marston criticized bungling operators, asserting that of more than 3,000 supposedly vaccinated persons with smallpox that he examined, only 268 bore the marks of a proper vaccination. The problem, he said, was the frequent failure to check, one week later, that the vaccine had "taken" and formed a proper blister. More significantly, it soon emerged that an epidemic of chickenpox was also in progress in Edinburgh, and these infections had been included in the smallpox figures. The problem highlights the uncertainty that surrounded distinctions between infections in the nineteenth century— some doctors were convinced that chickenpox was a form of smallpox.

Ever present in the arguments over vaccination was the burden of belief that a single vaccination gave lifelong protection. Jenner, wrongly, always believed this, and so did his adherents; it continued to be a source of trouble and doubt, undermining perceptions of the value of vaccination. The controversy continued for a hundred years. The *British Medical Journal* in 1896 concluded that single (Jennerian) vaccination was vastly successful in reducing death from smallpox but that it also postponed the risk of death from childhood into adult life. By replacing the permanent protection given by the real disease with a single infant vaccination of waning power, the practice resulted in a vast saving of life, but a shift of mortality into adult life. Modern recommendations on smallpox vaccination in the late twentieth century stipulated a booster vaccination at three- to ten-year intervals.

Following the severe smallpox epidemics in Paris in 1822 and 1825, in which more than 4,000 people died, a campaign for revaccination to put an end to repeated visitations was launched. The German kingdom of Württemberg was the first to introduce revaccination in 1829, and five years later it was made compulsory in the Prussian army. Russia, Denmark, and other states in Germany soon followed, but the British army waited until 1858 before making revaccination obligatory.

That smallpox could ravage London in the mid-nineteenth century, despite the existence of a new medical measure to prevent it, was a source of great frustration to the vaccinationists. In 1840 the epidemiologist William Farr railed against the fact that five children a day were dying of smallpox in London. "Imagine the outrage," he wrote in the *Lancet,* "if five children a day were thrown off London Bridge; the smallpox deaths could be stopped in a week if every unvaccinated person in London was vaccinated."

The Vaccination Act of 1840, which outlawed variolation in the United Kingdom, also introduced free vaccinations for the poor. Responsibility for enactment was given to the Poor Law Commissioners; they were to employ vaccinators at one shilling and sixpence per operation to be paid for out of the rates. The system was very unpopular for three reasons: the poor already hated the commissioners because of the stigma and disgrace of receiving their charity; doctors disliked working through the Poor Laws and distrusted their methods; and one-and-sixpence was a derisory sum, unlikely to attract good practitioners. Despite the problems, few favored compulsory vaccination, believing fines and punishments were best kept out of this humanitarian effort.

For the next ten years, two-thirds of smallpox deaths continued to occur in children. What more could be done? In 1853 the British government enacted more legislation, this time for the vaccination of all

children by four months of age with punishments for parents who refused. But the responsibility still lay with the despised Poor Law Commissioners, and threats of prison made the whole scheme unpopular.

Although the acts of 1840 and 1853 were partially successful (they
both halved the smallpox death rate), there were still compelling reasons to make the legislation work better by standardizing vaccination
practice and inspecting operators (anyone was allowed to vaccinate,
including the rude practitioners of folk medicine). In many European
countries, vaccination was better organized than in Britain; vaccination
certificates were often required for schooling, providing a strong incentive without fines or threats of imprisonment.

But things were worse in Scotland. Smallpox was rife among the
urban children of Lothian and Strathclyde. Imagine what it was like to
witness this, knowing an antidote was available. No wonder the professor of medicine at Edinburgh despaired of the ineffective programs in
Scotland when he could see the excellent achievements in Europe, and
even in the far-flung colonies, including the "savages of new Zealand."

While great advances in practice took place in many regions of the
globe, smallpox remained a scourge for native peoples of the Americas
throughout the nineteenth century. For the tribes of North America, vaccination policy was too bound up with displacement and loss of their
native lands to be welcomed. As many as 300,000 died in the epidemics
of 1836–40 in California, and things were no better for the indigenous
peoples of the southern continent. In Mexico, Argentina, Chile, Brazil,
and other great nations, the problem of organizing vaccination for native and rural populations remained enormous. But at least there was a
growing realization of what *could* be achieved when campaigns were
properly conducted. In Puerto Rico, annexation by the United States
in 1898 brought widespread improvements to infrastructure and public
health services, enabling the eradication of smallpox, through vaccination, within a year. Other countries struggled to follow suit. Across the
globe, governments, civic authorities, and national administrations began to share the vision of a world where smallpox could be controlled
and perhaps one day defeated.

The single most difficult aspect of spreading vaccination across the
globe was the fragile nature of the human vaccine chain, which depended
on arm-to-arm transmission. In some ways this worked quite well; vaccinators could insist on checking that the vaccine had "taken" five or
six days later, and only then issue a certificate and also take fresh lymph
to be passed on. However, there were two sets of problems. First, the
logistic difficulties of maintaining these "warm chains" meant that supplies were always limited and progress slow. Second, there was a serious
hazard of transferring chance infections. Among the latter was syphilis.

During the American Civil War (1861–65), vaccine was in very short supply, and desperate soldiers began vaccinating themselves arm to arm, believing that the bigger the vaccine wound, the greater the protection. Both syphilis and tetanus were spread in these uncontrolled and insanitary conditions, leading to rumors about the perils of the process just when war was favoring the spread of smallpox. In Europe too, transmission of syphilis did occur, partly because infants infected with syphilis in the womb showed no signs of their infection and were therefore included in arm-to-arm vaccination chains. In Rivalta, Italy, in 1861, 45 children were infected by a single syphilitic donor. Fifty other cases were documented in Europe between 1800 and 1880, leading to an estimated 750 cases of vaccinal syphilis. Although this was a tiny proportion of the total (around 100 million vaccinations), it disproportionately affected perceptions. Another bacterial infection, erysipelas, was a serious problem when vaccinations were done carelessly. In one severe outbreak of this aggressive streptococcal skin infection in England in 1876, six vaccinated children died. Jaundice, caused by hepatitis virus transferred during vaccination, was another significant problem.

Both supply restrictions and the inadvertent transmission of disease could be avoided if arm-to-arm vaccination could somehow be superseded by a better method. The answer, once again, proved to be the cow. It may seem obvious to us now, but it's important to remember that the nature, specific identity, and relative immutability of infectious agents was shrouded in mystery. Resorting directly to the animal for vaccine also recalled the bestial nature of the primary source. Cowpox was a rare disease of cows, but to overcome this difficulty a Neapolitan man named Troja had the idea of inoculating a cow with human vaccine. Another practitioner, Galbiati, took on the technique. In 1840 the procedure was adopted by a third Neapolitan worker named Negri, who began transferring cowpox from cow to cow. Gradually, against considerable prejudice and opposition, the method became accepted.

This so-called "animal vaccine" received great attention when it came to light during discussion at an international medical congress in Lyon in 1864. As a result a French doctor named Lanoix visited Naples to learn more. He was so impressed by the advantages of abundant vaccine supply and the avoidance of human disease transmission that he at once purchased a calf, had it vaccinated, and boarded the train for Paris without delay, lodging the calf comfortably in the guard's van. In Lyon he stopped to vaccinate some children and transfer lymph to another calf, then he set off again for Paris. His colleague Dr. Chambon installed the calf at his own house in the Saint Mandé, a suburb of Paris; out of this bold and decisive initiative grew the Paris Institute of Animal Vaccine, an inspirational model for many other institutions across Europe.

Here, sanitary aseptic techniques were developed to improve the quality of vaccine. In Belgium the National Vaccination Institute was established in Brussels in the Botanical Gardens in 1868. The success of "animal vaccine" was gradual. Traditional arm-to-arm practices remained, not least in the French Academy of Medicine, which only accepted the new practice after years of argument and bitter dispute.

Because the freshness of vaccine was important (the French had early disappointments with spoiled material), calves or heifers (young female cows) were enthusiastically transported to places where vaccination was needed. Such measures were eventually adopted by French vessels destined for smallpox-afflicted regions. A vaccinated heifer, her flank covered in repeated vaccination sites, was often included in the passenger list. In America, a Boston entrepreneur, Dr. Henri Matin, received vaccine from three French heifers in September 1870. He went on to propagate good-quality animal vaccine, eventually establishing himself as supplier to the whole of the U.S. East Coast.

In Britain tradition held sway—animal material was not widely used until 1880, when it became established alongside the original Jennerian method. In Italy, the pioneer of this method, the great majority of vaccine by this time was already from the animal source. The arm-to-arm procedure continued in Britain until the Vaccination Act of 1898 finally made it illegal.

One crucial aspect of vaccine source was that each time a new stock was introduced, uncertainty surrounded its quality and effectiveness until these things had been thoroughly tested. With lines of vaccine established in heifers on a manufacturing scale, controlling quality became a little easier, despite the invisible nature of the important component. Producers used only healthy animals with no signs of disease, maintained in clean conditions. Because the true nature of the infectious agent was unknown, the relationship between different vaccines could never be found out. Tried and tested was the only yardstick.

The steady progress in vaccine production was very much needed, because the battle against smallpox suffered massive swings of fortune in the nineteenth century. In France in 1870 an outbreak of smallpox had just begun, and a third of the population was unvaccinated when war broke out with Prussia. The massive movements of troops and the exodus of thousands from Paris quickly carried the French smallpox outbreak through the land. With the breakdown of organized measures to treat the sick and contain the infection, the disaster was compounded. Smallpox killed between 60,000 and 90,000 French, including more than 20,000 soldiers. The transportation of French prisoners of war also spread the contagion to the German states. The German army, more than a million strong, well-vaccinated and revaccinated, lost less than 500 men to smallpox, but civilians were less protected—more than 160,000 Germans died. French refugees carried the sickness to the Netherlands, Britain, and Sweden, while French troops took it to Switzerland. Tens of thousands died in Austria, and immigrants carried it across the ocean to the far shores of America. In England the death toll from smallpox rocketed to more than 40,000.

The armed conflict itself was soon over. German forces won a decisive battle at Sedan in northeast France, where they captured Napoleon III, in September 1870. But across Europe, more than a half-million died of smallpox. In the year 1871, the speckled monster had no better ally than crowded cities and a state of war.

And yet, despite these terrible events, the progress of vaccination in the nineteenth century was such that even the Franco-Prussian War was little more than a setback in the battle against smallpox. The figures for smallpox deaths bear this out. Defining an epidemic year as one in which a tenth of all deaths are caused by smallpox, then looking at London in the seventeenth century, 10 of the 48 years in which records were made qualify as epidemic years. In the eighteenth century, as the scourge of smallpox increased, there were 32 epidemic years. But in the nineteenth century, not even 1871 would qualify as an epidemic year. At last it had begun to seem as if Jenner's dream of eradicating smallpox might yet be realized.

8

A GREAT AND LOUD
COMMOTION

He [Killick Millard, Medical Officer of Health for Leicester] *gave a
practical demonstration of his faith in the power of vaccination by
taking his wife and two young children—all recently vaccinated—
into the smallpox hospital and photographing them by the bedside
of a patient with severe confluent disease.*

From a *British Medical Journal* obituary, 1952

One hundred years after Jenner's discovery, the immense success of vaccination was undeniable. Rates of smallpox infection fell tenfold when vaccination was introduced, another threefold when compulsion came in, and approached zero when revaccination was instituted. Success was overwhelmingly obvious.

But the very success of vaccination jeopardized support for continued campaigns as a safeguard against resurgence. Like any medical procedure, vaccination was by no means completely safe. The vaccine contained a living virus whose growth was readily limited by the immune system. But in the tiny minority of children with suppressed or damaged immune systems, it could grow uncontrolled and ultimately lead to encephalitis and death. Jenner himself had recognized the dangers of vaccinating children who had herpetic skin disease or severe eczema because the cowpox "virus" could become invasive. Because of the nature of the vaccine source, it was difficult to avoid the dangers of bacterial infection at the vaccination site; such infections could have serious consequences. The addition of glycerol killed some but not all types of bacteria, and it wasn't until the addition of phenol in the twentieth century that this risk was eliminated. In 1896 the feminist reformer (and the first female doctor in England) Elizabeth Garret Anderson reviewed the Royal Commission's report and estimated a figure of 50 deaths per year from vaccination in England and Wales. This meant a fatality resulting from approximately one in every 14,000 vaccinations.

Recalling the historical sequence of risk: smallpox brought a 1 in 5 chance of death (and certain disfigurement); early variolation carried a 1 in 50 chance of proving fatal; refined variolation at its best had a 1 in a 1,000 chance; and now Jennerian vaccination, in the age of antiseptic medicine, had at most a 1 in 14,000 chance of causing death. This was highly laudable in the context of the historical improvement, but with the dangers of smallpox ever more remote, it was enough to stir up anti-vaccination sentiment.

By far the biggest cause of resentment in Britain was compulsion. In 1853 vaccination was made compulsory in England and Wales, and this provoked the formation of the Anti-Vaccination League. A similar organization was later founded in the United States. In 1863 compulsion in England and Wales was extended to Ireland and Scotland, and the act was strengthened four years later with the appointment of officers to enforce the law. One by one the nations of Europe followed suit. In France, public policy required the direct animal source; Germany passed stringent laws in 1874, which included revaccination. In Switzerland compulsion laws were framed in 1882 but defeated by a referendum of the people. Ten years later, Italy made vaccination compulsory, and Holland made it a prerequisite for attending school.

The opponents of vaccination have sometimes been surprising. Alfred Russell Wallace was a gifted naturalist, explorer, and geographer, and the brilliant originator of an observation-based theory of evolution that prompted Darwin to publish his great work on the origin of species. Wallace claimed that the success of vaccination was founded on "useless and dangerous statistics," fervently asserting that "vaccination is quite powerless either to prevent or mitigate smallpox." He went on to condemn vaccination as "an operation which had admittedly caused many deaths, which is probably the cause of greater mortality than smallpox itself; but which cannot be proved ever to have saved a single life." Darwin, by contrast, supported vaccination.

Another distinguished critic of vaccination was George Bernard Shaw, who denounced vaccination as "a particularly filthy piece of witchcraft." Perhaps Shaw's choice of words was colored by his zealous vegetarian beliefs: "A man of my spiritual intensity does not eat corpses." But the British opposition to vaccination had its deepest and most ardent roots in the ancient city of Leicester in the heart of England.

In 1885 this growing, recently industrialized town on the River Soar had around 100,000 inhabitants. Of these, an extraordinary 4,000 were awaiting prosecution for failing to vaccinate their children. A protest meeting against compulsion and punishment quickly grew to 20,000 strong, and a copy of the Vaccination Act was ceremonially burned. Riots broke out and the crowds made "a great and loud commotion."

In that year, a new medical officer of health for Leicester, Joseph Priestly, was appointed. He was a staunch supporter of vaccination, but the town's Sanitary Committee overruled his attempts to promote the practice. Interestingly, the very strong sentiments in Leicester were not based on irrational fears and illogical beliefs but on faith in an alternative: the strict practice of hygiene and the efficient identification and isolation of smallpox cases, which was championed by the Leicester hygienist J. F. Biggs. Biggs later wrote a lengthy book entitled *Leicester: Sanitation Versus Vaccination*. The law on vaccination was still administered by the Poor Law Guardians so the inventive people of Leicester elected guardians opposed to vaccination. Their strategy was a success. The prosecutions ceased, and the child vaccination rate, which reached 90 percent in 1872, fell dramatically to 3 percent by 1892.

The town of Leicester, with almost all its children going unvaccinated, seemed to be asking for trouble. Because it had been free of smallpox for many years, and because natural immunity caused by infection was at an all-time low in the region, any outbreak of smallpox would quickly run through the crowded city and wreak havoc among the unprotected population. It was only a matter of time.

But for Leicester, such a fate was not to be. Priestley's replacement was a man named Killick Millard, and he tried a different tack. Millard was against compulsory infant vaccination, believing instead that when an outbreak occurred, vaccination of all contacts was sufficient. He avoided confrontation by promoting sanitation, disinfection, and vigilance together with vaccination used strategically in this way. His enlightened strategy helped mitigate the vicious way in which vaccination was enforced in the supremely sensitive area of parental concern for their children's well-being. Compulsion was seen as an infringement of individual rights and self-determination; it could also lead to sloppy vaccination practice, and it did great damage to vaccination policy. As a result of Killick Millard's thoughtful approach, no catastrophic smallpox epidemic overtook the vulnerable town. The "Leicester Method" would eventually come to mean the notification, isolation, and disinfection measures so ardently supported by Biggs and his fellow citizens, together with surveillance and prompt vaccination of contacts.

In the same decade, the cathedral city of Gloucester (Jenner's county town) had a much less fortunate experience. After twenty years of freedom from smallpox, the city was without provision for emergency. Antivaccinationists, supported by the local press, had been at work, and in

1895 a quarter of the population went unvaccinated. When smallpox broke out it quickly claimed 2,000 victims. More than 400 perished. Smallpox spread quickly among unvaccinated children, hospitals were overrun, and all attempts at quarantine broke down. The whole city fell into confusion as trade and business faltered.

The original misconceived belief—still held by some practitioners and many ordinary people, that a single vaccination gave lifetime protection against smallpox—was a crucial cause of Gloucester's woes and the single most significant obstacle to public acceptance of vaccination. Both sides in the vaccination debate were guilty of distorting facts and figures, and provaccinationists denied that infection was possible in those properly vaccinated. The reality in Gloucester, and elsewhere, was different. Mild infections in adults vaccinated in childhood went unnoticed, and these spread smallpox to the unprotected. Drastic measures were needed to bring the town's disaster to an end. Compulsory vaccination and revaccination was enforced by six vaccinators who traveled house to house until the entire population had been treated and the disease had been expunged at last.

Despite such harsh lessons about the need for revaccination, the measure was never enforced through compulsion in Britain. The Royal Commission—set up in 1885 in response to the Leicester riots—concluded (after seven long years of deliberation) that "revaccination restores the protection which lapse of time has diminished." But the commissioners knew it would be counterproductive if they tried to bring it into law.

Instead the commission swung the other way, recommending the abolition of cumulative penalties for failing to vaccinate children. Parliament responded with the 1898 Vaccination Act, which introduced a conscience clause: although vaccination remained compulsory, parents who didn't believe vaccination was effective or safe for their children could apply for exemption. The militant Leicester protesters and the antivaccinationists had finally achieved their goals, but it was also a victory for common sense. Compulsion for all, when it was enforced by fines and imprisonment, was a source of social strife. A quarter-million exemption certificates were granted in the first year, and the wider concept of conscientious objection became established for the first time.

Experts predicted disaster. The ravages of smallpox would surely return, they said, if half the people didn't vaccinate their children. A cartoon in the British satirical magazine *Punch* portrayed the "triumph of De-Jenner-ation": the skeletal figure of the Grim Reaper, his dark cloak billowing behind him, holds aloft a scrolled copy of the Vaccination Act, a dreadful grin of jubilation on his face. Behind him, a serpent wakens from its sleep and an hourglass, upset from a broken pedestal, measures out the hours till death.

But somehow the prophesies of learned men proved wrong and no catastrophic epidemic materialized. The reasons for this probably have to do with the mysterious nature of smallpox. The severe form of smallpox (which we now know is caused by *Variola major*) was dominant in the eighteenth century. The other, much milder version (caused by the viral variant, *Variola minor*) was more commonly found in West Africa and the Americas; but around the time when exemption from vaccination was introduced in Britain, *Variola minor* was becoming more widespread. The reasons for this remain mysterious, but it seems likely that the combination of partial vaccination and revaccination, and the natural immunity arising from the spread of the mild variant of smallpox, meant that smallpox was no longer the scourge it had been. Because the mild form is much less debilitating, those affected tend to remain active and quickly transmit the infection, providing a natural source of immunity that spreads through the population without serious illness. This is what happened early in the twentieth century, when *Variola minor* spread through America, Great Britain, and South Africa to become the dominant form of the disease. In Britain, when the National Health Service was launched in 1948, smallpox vaccination was no longer compulsory. Routine vaccination ended in 1971, 173 years after Edward Jenner's first experiment.

9

COMPLETING
THE PICTURE

"If when hearing that I have been stilled at last, they stand at the
*   door,*
Watching the full-starred heavens that winter sees,
Will this thought rise on those who will meet my face no more,
'He was one who had an eye for such mysteries'?"

Thomas Hardy, 1917

What would Jenner think of the "brickbats and hostile weapons of every sort" wielded against vaccination in the 1890s? He probably wouldn't have been too surprised. These, after all, were his own words a hundred years earlier. And he well knew that aspect of human nature that ensures no one wants to be bothered with the least unpleasantness when nothing serious seems to threaten. In 1814, one year after Oxford University officially honored him with the degree of Doctor of Medicine, Jenner made what was to be his last visit to London.

It was an interesting year in the capital. The last frost fair to be held in London took place early in February, and just below Blackfriars Bridge an elephant was ceremoniously led across the frozen Thames. Gas lighting was introduced on the streets, and printing underwent a revolution when the *Times* newspaper adopted steam power. By now King George III, Jenner's great patron, had become unfit to rule. His illness, a genetic defect in body chemistry, had led to insanity, and his son, the Prince Regent (later, George IV), ruled in his stead. It was a time of elegance and achievement in the fine arts and architecture. Upper class society flourished in a sort of mini-Renaissance of culture and refinement, and no expense was spared on celebrations when, in April, the sovereigns and generals of the Coalition Allies—Austria, Prussia, Russia, Sweden, and other German states—visited London to celebrate the peace after Napoleon's defeat. The peace was not entirely complete—Britain was at war with the United States, and in August, British troops in Washington would burn the White House.

Among the visiting heads of state was Tsar Alexander of Russia, King Frederick William III of Prussia, and Prince Metternich, chancellor of the Austrian Empire. After many parties, banquets, and parades, the assembled monarchs and marshals attended the June horse races at Ascot and then visited Oxford, where Tsar Alexander, King Frederick William, and Marshal Blucher, like Jenner the year before, were honored with degrees. Jenner himself was presented to the august assembly at the magnificent palace of Carlton House in St. James.

But Jenner didn't tarry too long in London. There was a dark side to the beauty, high fashion, and extravagance of the upper classes, and London was more commonly a place of dirt and noise. The population of London was growing fast, and according to the scholar and poet Robert Southey, the difference between the strata of society was vast indeed: "The inhabitants of this great city," he said, "seem to be divided into two distinct casts,—the Solar and the Lunar races."

The lowest of the Lunar races lived in slums or "rookeries," or else survived unsheltered on the streets. Among the worst of them was "Rats Castle" in St. Giles, a mile from the pomp and splendor of Carlton House and two miles north of Westminster Bridge. Here the poor dwelled in deplorable conditions. Low-quality housing was flung up wherever there was free space, and a single room might house an entire family, or more than one family. Two years after Jenner's visit, a Parliamentary Committee was set up to look into the problems of the London slums and what might be done about them. Professionals were called to give evidence of their experience. One London doctor, William Blair, had this to say:

> Human beings, hogs, and dogs, were associated in the same habitations; and great heaps of dirt, in different quarters, may be found piled up in the streets. Another reason of their ill health is this, that some of the lower inhabitations have neither windows nor chimneys nor floors, and are so dark that I can scarcely see there at midday without a candle. I have actually gone into a ground floor bedroom, and could not find my patient without the light of a candle.

Jenner, who was always concerned for the plight of the poor and ministered to them whenever he could, would have known this side of London all too well. Quite soon after his audience with the rulers of

Europe, he returned to his beloved Gloucestershire and helped organize a philosophical and literary society, reassuming his role as "Vaccine Clerk to the world." He advised and researched on the safest ways to produce and transport his vaccine and tried to sort out apparent failures of vaccination, which were usually due to outdated material or incompetent vaccinators.

Never truly happy in London, Jenner was likewise never truly comfortable (except in his student years) among its medical establishment. The fact that he visited London every year between 1798 and 1814 and spent four years (1801–1805) in practice near Park Lane is a measure of how determined he was to champion vaccination. His physical and figurative separation from the London scene is delightfully illustrated by the strange case of the Medical Society of London Portrait of Members. The society's founder, John Coakley Lettsome, a friend and supporter of Jenner, commissioned a portrait of the fellows—physicians, surgeons, and apothecaries—in 1801. This learned and progressive assembly sat for artist Samuel Medley, and an engraving was produced showing the members gathered in a Georgian library around an imposing desk draped with an oriental rug, with bookcases and library busts in the background. Lettsome himself presides at the center. Jenner, of course, was not present, but Lettsome realized that the portrait simply *must* include him. The figure of Jenner was therefore added afterward. With benefit of this knowledge, the late addition is all too obvious: the pale, almost ghostlike figure of Jenner occupies a notional space beyond and apart from the darkly solid circle of members, yet somehow he is prominent within the picture—a perfect metaphor for Jenner's relationship with the London medical establishment.

In 1810 Jenner's eldest son, Edward, died of tuberculosis at the age of 21. Another tragedy was soon to follow. On September 13, 1815, Jenner's wife, Catherine, who had been unwell for many years, succumbed at last to the same relentless disease. Jenner never left Berkeley again, except for a day or two, for as long as he lived. He immersed himself in studies relating to his great discovery and in his tireless work as a physician, naturalist, and magistrate. The plight of the poor and the rising level of crime continued to trouble him, and he maintained his practice of free vaccination in the "Temple of Vaccinia," an ornate timber summerhouse in his garden.

He kept a productive kitchen garden and a lovely ornamental garden enclosed by trees, and he installed an attractive hothouse at the rear of Chantry Cottage for the latest varieties of tender plants, importing seeds from Italy and Spain. He became expert at propagating fruit bushes such as gooseberries, raspberries, and figs. In 1818 he introduced young grapevines from the famous stock at Hampton Court.

For a man like Jenner, his naturalist's curiosity would always be endless and never truly satisfied. He rediscovered his passion for fossil hunting, searching out trilobites and crinoids (sea lilies) in the ancient beds along the Severn and the Cotswold Scarp. In 1819 he uncovered the fossilized remains of a plesiosaur below Stinchcombe Hill, a prehistoric marine reptile from more than 65 million years ago and the first of its kind to be discovered in Britain. Jenner called it a monument to the departed world. In 1822 he completed a paper, "Some Observations on the Migration of Birds," for submission to the Royal Society; his nephew George would present it posthumously in 1823. George, who helped with the study, apologized on behalf of his "revered Uncle" for not delivering the study earlier, "in consequence of his extensive correspondence with almost every part of the globe on the interesting subject of Vaccination."

In 1820 Jenner had a mild stroke from which he recovered well. Three years later, in January 1823 at the age of 73, he had another. The weather was cold, and he had walked to a neighboring village where he'd ordered fuel to be provided for certain poor families. Later he was called out to a friend and colleague who had suffered a stroke, doing his best for him but judging that his patient would not last another 24 hours. Next morning, Saturday, January 25, Jenner failed to appear at breakfast. A servant was sent to search for him and found him unconscious, stretched out on the floor of the library. His nephew Henry anxiously sought help, but to no avail; he then bled his uncle several times. Jenner never regained consciousness. His friend, John Baron, was urgently summoned from Gloucester, and James Phipps (then aged 35) hurried to Chantry Cottage before fetching another of Jenner's nephews, William Davies. Jenner had long ago provided Phipps with a house nearby when Phipps had fallen on hard times. Among the objects about his person was a dried rose Catherine had given him and which Jenner carried with him. During the early hours of Sunday, January 26, 1823, Edward Jenner died. His last patient outlived him by a few hours.

Jenner was buried in the family tomb beside the altar in Berkeley Church, next to his parents, his eldest son, and his wife. John Baron wrote an obituary, quoted throughout Britain in leading journals and newspapers, that described the funeral ceremony: "The concourse of persons was immense; the indications of respect, reverence and regret were unequivocally conspicuous; every eye was moistened; and every heart oppressed." But in reality the funeral remained a local affair, with no one attending from London. Jenner was survived by his son Robert and his daughter, Catherine. An inscription intended for his tomb and appended to many obituaries was never engraved. It is a verse that expresses the emotions attending his death, and I think he might have enjoyed it. It would certainly have riled his enemies.

Within this tomb hath found a resting place,
The great physician of the human race—
Immortal Jenner! whose gigantic mind
Brought life and health to more than half mankind.
Let rescued infancy his worth proclaim,
And lisp out blessings on his honoured name;
And radiant beauty drop her saddest tear,
For beauty's truest, trustiest friend lies here!

Personally, I feel the man is better served by the poem that opens this chapter, but unfortunately another verse in Thomas Hardy's poem clearly disqualifies it.

If I pass during some nocturnal blackness, mothy and warm,
When the hedgehog travels furtively over the lawn,
One may say, "He strove that such innocent creatures should come to
    no harm."

So perhaps another inscription must serve instead. It is the one that decorates Jenner's statue in Kensington Gardens and reads: JENNER.

10

GERM THEORY AND THE BIRTH OF IMMUNOLOGY

"In the fields of observation chance favors only the prepared mind."

Louis Pasteur, 1854

I n many ways it's hard for us to grasp just how little was known about the nature of germs, microbes, or viruses throughout Jenner's lifetime. Perhaps his greatest insights had to do with understanding the properties of such invisible agents, different from each other and causing different diseases but also interrelated. Crucially, he used these insights to practical ends without too much speculation. But others had plenty of ideas about what smallpox "germs" might be like.

Two centuries earlier, in 1546, the physician, scholar, and poet Girolamo Fracastoro had proposed a concept brilliantly foreshadowing germ theory. Fracastoro was born in 1483 and grew up in the city of Verona in the northern plains of Italy, surrounded by its Roman riches: pink and white limestone, the loggias and campaniles of churches, palaces and theaters in the loop of the Adige River. The river hurries on its way, but the city keeps its treasures: today the modern shopping streets are paved in pink marble, and Fracastoro, too, remains—a marble figure on an archway in the medieval Piazza dei Signori, ringed by shadowed cloisters.

Fracastoro was born to an upper-class family and was something of a prodigy. At the tender age of nineteen, he was appointed to the chair of logic and philosophy at the nearby University of Padua. He is best remembered for two great works on medicine. The first provided a description, and an unambiguous name, for a new disease in Europe traditionally blamed on despised rival nations—the "French disease" in Italy, the "Italian disease" in France, and so on. Fracastoro's scholarly description took the form of an epic poem in which the protagonist is a shepherd boy on a Caribbean island who offends the sun god and whose people are duly punished with disease. The shepherd's name is Syphilus, and the island people soon pass on the infection to Spanish sailors who in turn carry the terrible affliction to Europe. The theory that syphilis was brought to Europe by seamen returning with Columbus remains a favored explanation. Fracastoro's work, entitled *Syphilidis sive Morbi Gallici* [*Syphilis, or the French disease*], was published in 1530. The text provided a graphic description of the secondary and tertiary phases of syphilis; it recognized the sexual transmission of infection and suggested treatment with a plant extract, oil of guaiac.

In his second great work, in prose this time, *De Contagionibus et Contagiosis Morbis et Earum Curatione* [*On Contagion and Contagious Diseases,* 1546], Fracastoro elucidated his theory of the nature of infection. In this work he attributed epidemic diseases to agents too small to be seen, transmitted between people either by direct contact or indirectly. However, he did not characterize these invisible spores as living organisms. He called contaminated clothes, bedding, and similar articles *fomites* (from the Latin *fomes,* meaning tinder), describing them as not in themselves corrupt, but capable of fostering the essential seeds of the contagion. Fracastoro viewed both smallpox and the plague as contagious. Verona was repeatedly afflicted by plague in his lifetime; in Shakespeare's *Romeo and Juliet,* when Mercutio says, "A plague on both your houses," his curse on the Montagues and Capulets is very much to the point. Fracastoro was also an astronomer and complemented his theory of contagion with notions of the influence of the stars on the spread of diseases.

But in later centuries, Fracastoro's brilliant forerunner to germ theory wasn't much used in thinking about smallpox. Cotton Mather, the charismatic puritan preacher who championed the use of variolation in Boston in 1721, called smallpox a "miasmic enemy" (like the poisons in foul air) invading the citadel of the body. He believed the enemy fed on some vital body substance, using it all up. This explained immunity as some kind of depletion of essential fuel, thus a particular kind of germ could never strike the same individual twice. Variolation, or immunization, used up this substance with only mild symptoms of disease. A passing suggestion about germs and immunity, much closer to what actually happens, was made by James Kirkpatrick, a physician from Charleston, South Carolina, in 1754. He speculated that people who've had smallpox must have something remaining in their body that could overcome a subsequent infection.

But there are two mysteries here, intimately entwined. The first is the true nature of "germs"; the second is the nature of immunity—how the body fights against them. Neither was understood in Jenner's time, but there were clues and ingenious ideas about both.

The first sighting of bacteria under the microscope was made by Antoni van Leeuwenhoek in 1676, when he was investigating the taste of foods. He described them as much smaller than the single-celled animals (protozoa), which he had already seen in lake water. Born in Delft in the Netherlands in 1632, the same year (and month) as the painter Jan Vermeer, Van Leeuwenhoek was the son of a basket maker, so he did not enjoy a privileged education. Instead, he was apprenticed to his uncle in a linen-draper's shop in a nearby town and returned to Delft in 1654, where he established himself as a draper. He also worked as a surveyor, wine assayer, and minor city official.

In 1676 he served as a trustee of the estate of the deceased and bankrupt Vermeer, and some speculate that the two men were friends. The notion is attractive because of Vermeer's precise use of perspective to create the serene stillness of his domestic interiors. These painted spaces, which brim with a gentle light, seem filled, in the words of art historian David Piper, "with the silence of unheard music." But in order to create these perfect compositions, Vermeer may have used instruments of perspective akin to the *camera obscura,* and the preeminent pioneer of optical instruments was Antoni van Leeuwenhoek, who lived a few streets away in the small Dutch town. It's nice to imagine Van Leeuwenhoek at work in one of the tranquil chambers painted by Vermeer, or walking the canal-side streets beneath stepped gables and red-tiled rooftops bathed in the soft radiance of cloud light that Vermeer so loved.

Van Leeuwenhoek's supremacy as an optical engineer was not just of interest for the new science of perspective. In 1673 Van Leeuwenhoek began writing letters to the newly formed Royal Society of London, describing what he had seen through the powerful lenses that he hand-crafted. For 50 years he corresponded with the Royal Society, describing microscopic life. "In structure these little animals were fashioned like a bell, and at the round opening they made such a stir, that the particles in the water thereabout were set in motion thereby. And though I must have seen quite 20 of these little animals on their long tails alongside one another very gently moving, with outstretched bodies and straightened-out tails; yet in an instant . . . they pulled their bodies and their tails together, . . . [before] continuing their gentle motion: which sight I found mightily diverting." Van Leeuwenhoek went on to describe protozoa and bacteria in human feces from infected patients, although he didn't speculate they might be causing the illness.

Another Dutchman, Herman Boerhaave, born in Leiden in 1668, argued that smallpox was caused by a contagious agent. Boerhaave saw no reason to invoke arcane complexities when the truth was often so much simpler, and he sought to counter the belief that diseases such as smallpox were caused by miasmic influence. He is often regarded as the founder of clinical teaching and the modern academic hospital. Before this time, while it was understood through observation that diseases were infectious, there were no clear ideas about the nature of infectious agents; scientific and medical publications blamed poorly defined causes such as the ethereal influence of swamps, sewers, and noxious atmospheres.

Throughout Jenner's lifetime, developments in both microscopes and ideas about infection led to links being made between microorganisms and diseases, but understanding didn't really accelerate until the 1870s. An important step forward was the realization that bacteria and other "germs" of decay did not arise spontaneously from nonliving matter. In 1859, the French Academy of Sciences sponsored a contest for the best experiment proving or disproving spontaneous generation. The winner was a young chemist named Louis Pasteur.

Pasteur was born on December 27, 1822, in Dole in the Jura region of eastern France. His father was a tanner, and the family lived in Canal des Tanneurs, whose cellars looked onto the canal and tanning pits. Above this was the shop and workshop, with the drying shed at the top and the family living between. Like Jenner, Pasteur's house is now a museum to his achievements. In school he was an unremarkable student, but his talents were eventually recognized. He studied at the nearby Collège Royal de Besançon, and in 1847 he earned a doctorate from the École Normale in Paris. After several years of research and teaching in Dijon and Strasbourg, Pasteur was appointed professor of chemistry at the University of Lille. The official duties of the faculty of sciences included searching for solutions to the practical problems of local industries, particularly vinegar production and winemaking. Pasteur was able to demonstrate that microorganisms were responsible for souring vinegar, wine, and beer, and later milk, and that these could be killed by boiling and then cooling. (The modern process of pasteurization uses a temperature of at least 72°C for sixteen seconds.)

But proving where these microbes came from would be a momentous step forward. To win the French Academy's prize, Pasteur demonstrated that bacteria did not appear in meat broth boiled in a flask ingeniously shaped to let in air but not particles (it had an S-shaped neck which trapped particles through settling). Simply by tilting the flask, so any invisible particles came into contact with the broth, he showed the broth soon became cloudy with life. Pasteur had both refuted the theory of spontaneous generation and demonstrated that microorganisms are everywhere—even in the air. He then set about trying to discover how germs cause disease.

Bacteria could at least be seen, but this was not the case with viruses. It was not until 1892 that viruses were first defined in any way. The Russian biologist Dimitri Ivanovsky, working at the University of St. Petersburg, discovered that a certain disease of tobacco plants could pass through unglazed porcelain filters (these filters had the smallest pores available at that time) while bacteria could not. At the time, all infectious agents were thought to be too large to pass through the pores of such filters and to be able to grow on nutrient broth. In 1898, the Dutch microbiologist Martinus Beijerinck repeated the experiments and became convinced that the filtered solution contained a new form of infectious agent that could only thrive in growing plants—a liquid living thing that he called a *virus*. An Italian microbiologist, Adelchi Negri, showed that the smallpox germ was also a filterable agent, or virus, in 1906. It wasn't until much later that viruses were shown to exist as particles visible under an electron microscope. An American chemist, Wendell Stanley, of the Rockefeller Institute in New York City, established this for tobacco mosaic virus; he earned a Nobel Prize in 1946 for the discovery.

In 1865 Pasteur lost his two-year-old daughter to infection, and less than a year later, her twelve-year-old sister died of typhoid fever. Two years later, overworked and grief-stricken, Pasteur suffered a cerebral hemorrhage that left one arm and one leg partially paralyzed. But he didn't let this slow him down. In the summer of 1880, Pasteur was studying fowl cholera, keeping the infectious material in flasks and injecting it into chickens. One day a colleague couldn't find time to inject the cholera into a new group of chickens as planned. The flasks stood untended on the laboratory bench. When at last he got around to injecting the material, the chickens didn't die. Intrigued, Pasteur asked his colleague to repeat what he'd done but with a fresh culture of cholera. Two groups of chickens were inoculated: the ones that had already been given the old culture and ones that had not. The chickens that had been given the old culture survived while those that hadn't died. Pasteur quickly realized that the old weakened cultures had made the birds immune to cholera, a historic breakthrough in understanding the nature of immunity.

Pasteur, who knew all about Jenner's work, then tried to make a vaccine against anthrax. He thought that weakening the germ by exposing it to oxygen might do the trick, and he treated anthrax in a flask with a chemical oxidizing agent. In April 1881, Pasteur announced that his team had found a way to weaken anthrax germs and so could produce a vaccine against this terrible disease. Despite Pasteur's renown, or because of it, there were many in the medical establishment who mocked him. One of them, Hippolyte Rossignol, the editor of the *Veterinary Press,* challenged Pasteur to a public trial of his vaccine. Failure would have resulted in the humiliation of the famous man. In May 1882, at Pouilly-le-Fort near Fontainebleau, 25 sheep were inoculated with Pasteur's vaccine while another 25 were not. All 50 were then injected with the anthrax germ. Those that were not inoculated died within two days, but the inoculated group suffered no ill effects and was described as being "sound, and [they] frolicked and gave signs of perfect health." The *Times* in London called Pasteur "one of the scientific glories of France."

Pasteur then turned his attention to rabies. He grew the rabies virus in the spinal cords of rabbits, and when these were removed and stored for a couple of weeks, he found the germ was weakened. When he injected dogs with stored spinal-cord tissue containing the virus, they showed no sign of the disease. Before Pasteur could proceed to properly test the material as a vaccine, a young boy named Joseph Meister was brought to him. Joseph had been bitten by a rabid dog, and it seemed he would certainly die an agonizing death if nothing could be done for him. Pasteur took the risk of using his untested vaccine, saying, "The death of this child appearing to be inevitable, I decided, not without lively and sore anxiety, as may well be believed, to try upon Joseph Meister, the method which I had found constantly successful with dogs." Sixty hours after the rabid bite, he injected the vaccine (stored rabbit spinal cord tissue) under a fold of skin. He then made thirteen inoculations over many days using progressively younger cord material, the last of which contained highly virulent rabies virus. The boy remained perfectly well, and Pasteur realized he had indeed discovered a vaccine against rabies. Like Jenner's famous vaccine, no one had the slightest idea how it worked.

It was Pasteur who honored Jenner by applying the term vaccine to all inoculations against infectious disease. He subscribed, for a while at least, to the depletion theory of immunity, but there were other theories around, such as notions about microbes leaving something in the body that prevented the microbes from growing on a second visit. However, all these ideas assumed the body was somehow passive, and that it was the germ that actively used up something essential or left behind something poisonous to itself. These theories went out of fashion in the 1880s, when it was shown that substances present in normal blood could kill bacteria. Not long afterward, a spectacular discovery was made.

In 1890, in Berlin, two bacteriologists were working with toxins extracted from the newly identified bacteria that caused tetanus and diphtheria. One of them was Emil von Behring; the other was his Japanese student, Kitasato Shibasaburō. They injected small (nonlethal) amounts of toxin into guinea pigs and later harvested their blood. In the serum they found a mysterious substance produced in response to the toxin injection, which neutralized that particular toxin but not other toxins. Toxin mixed with this serum and then injected into fresh guinea pigs was found to be harmless. They called the new substance *antitoxin*. This was the discovery of antibody—the proteins actively produced by the body to neutralize, eliminate, or kill invading bacteria and viruses. In collaboration with Paul Ehrlich, a distinguished bacteriologist, Behring developed a diphtheria antibody with which he hoped to treat children already infected with diphtheria. When, in 1894, French investigators Émile Roux and Alexandre Yersin introduced the immunization of horses, Behring immediately adopted this procedure to make large quantities of antibody. He used this to inject children who had diphtheria in an attempt to neutralize the deadly toxin. The new treatment was immediately successful and resulted in a dramatic fall in mortality. In 1901 Emil von Behring received the first Nobel Prize for medicine for his work, which "placed in the hands of the physician a victorious weapon against illness and death."

With the discovery of antibodies, the science of immunology was born. Modern immunologists are a lively, friendly, highly combative group of scientists bent on describing, in the greatest detail, the immensely complicated, often surprising ways in which the body defends itself and learns to fight invaders. But the greatest, most rancorous, and bitterly divisive arguments in immunology took place very early in the history of the field. In the last decade of the nineteenth century, news was breaking almost every month of new microorganisms, new diseases, new mechanisms; in the fevered, inflammatory world of immunology, two passionate debates were raging. The first was about whether inflammation was an abnormal, harmful manifestation of disease or whether it was a defensive response to infection. The second concerned the nature of the important defenders: were they the invisible antibodies in blood and body fluids, or were they the circulating white cells whose behavior could be seen under the microscope? Views were entrenched, feelings ran high, and the arguments grew fiercer.

Behind the antibody school of thought lay 2,000 years of tradition—the other name used was "humoral immunity," directly linking it to the classical notion of vital humors. Behind the cellular theory lay a much shorter history. In 1859, Rudolph Virchow, a German physician and social reformer, proposed that underlying all human disease lay not a disorder of the humors but a disorder of the cells of the body. Virchow, an only child from a farming family in Schivelbein, Germany, received a scholarship from the Prussian Military Academy to study medicine in preparation for a career as an army physician. As an investigator, his great strength lay in meticulous and painstaking observation, including extensive use of the microscope. At the age of 24 he encountered a fatal illness in a middle-aged cook, and on examining her body he found an enlarged spleen and large amounts of white cell material usually associated with serious infections. But there was no sign of an infection. This prompted him to wonder if the blood itself was abnormal, and eschewing the meaningless, high-flown terms beloved of doctors at the time, he simply named it for its appearance: *white blood,* which he later rendered as *leukaemia.* He described the disorder, correctly, as a proliferation of the white cells. This observation-based description was the beginning of his cellular theory of the human body and the cellular basis of disease— a great leap forward in understanding. Despite his great efforts, most medical practices continued to be based on notions of invisible forces, including miasmas and imbalance of the four classical humors. Virchow went on to extend his studies to other cancers, investigating the dynamics of cancerous cells and developing an understanding of the uncontrolled and disordered nature of their growth. He coined the term we use still today for cancerous growth, *neoplasia.*

In 1884 a Jewish Russian biologist, studying the transparent young of starfish, observed the wandering white cells clustering around thorns that he inserted into their tissue. His name was Elie Metchnikoff, and his devoted second wife, Olga, afterward wrote a biography of her husband in which she renders the moment in his own words.

> One day when the whole family had gone to a circus to see some extraordinary performing apes, I remained alone with my microscope, observing the life in the mobile cells of a transparent star-fish larva, when a new thought suddenly flashed across my brain. It struck me that similar cells might serve in the defense of the organism against intruders. Feeling that there was in this something of surpassing interest, I felt so excited that I began striding up and down the room and even went to the seashore in order to collect my thoughts. I said to myself that, if my supposition was true, a splinter introduced into the body of a star-fish larva, devoid of blood-vessels or of a nervous system, should soon be surrounded by mobile cells as is to be observed in a man who runs a splinter into his finger. This was no sooner said than done. There was a small garden to our dwelling, in which we had a few days previously organized a "Christmas tree" for the children on a little tangerine tree; I fetched from it a few rose thorns and introduced them once at once under the skin of some beautiful star-fish larvae as transparent as water. I was too excited to sleep that night in the expectation of the result of my experiment, and very early the next morning I ascertained that it had fully succeeded. That experiment formed the basis of the phagocyte theory, to the development of which I devoted the next twenty-five years of my life.

Metchnikoff went on to establish the importance of phagocytic white cells in defense against disease: how they patrol the body, engulf invading microbes, and digest them. He showed that the inflammation provoked by an infection, when examined under the microscope, revealed large gatherings of white cells engulfing the invading pathogens. But he had to battle against the entrenched belief that inflammation was a bad thing.

In 1888 Metchnikoff joined the Pasteur Institute in Paris and effectively led the cellular school of thought, which was largely composed of French scientists. The humoralists were based in Germany, and prominent German pathologists lined up to attack the French theory that cells were important in immune defenses. When the German bacteriologist Richard Pfeiffer showed, in 1894, that the cholera germ was killed by serum from immunized animals that contained antibodies but not white cells, the Germans gained ascendency.

The conflict echoed the hostilities of the Franco-Prussian War and the bitter arguments between the patriotic Louis Pasteur and the great Prussian bacteriologist Robert Koch. Koch, who identified the bacteria that causes tuberculosis in 1882, was a formidable opponent. It was he who established the rigorous rules for proving that a particular disease is caused by a specific germ; his pupils identified the organisms responsible for diphtheria, typhoid, pneumonia, gonorrhea, leprosy, bubonic plague, tetanus, and syphilis using his methods. In his early career Koch resembled Edward Jenner: he was doing a routine medical practitioner's job when he made the discovery that would lead to fame. In 1870, age 27, Koch volunteered for service in the Franco-Prussian War, and from 1872 to 1880 he served as district medical officer for Wollstein near the Franco-Prussian border. Anthrax, at this time, was prevalent among local farm animals, and without any laboratory resources except his microscope, Koch carried out scientific investigations. The anthrax bacillus had first been observed under the microscope by the French physician Casimir Davaine and his colleagues in 1850. But Koch, isolated from the scientific community in the four-roomed flat that was his home, set about proving beyond doubt that this was the cause of the disease.

He took material from the spleens of farm animals dead of anthrax and injected it into mice using homemade slivers of wood. He found that the bacilli killed these mice, whereas mice injected with material from healthy animals did not suffer any disease. This confirmed what was already known, that anthrax could be transmitted with the blood of infected animals. But Koch was determined to go further. He wanted to find out whether anthrax bacilli never in contact with an animal could cause the disease. He managed to produce pure cultures of anthrax bacteria by growing them on the aqueous humor in the eye of an ox.

Koch drew and photographed these microbial cultures and recorded the growth of the bacilli, noting that when he altered the culture conditions to "starve" the bacteria, they turned into rounded spores that could resist the adverse conditions, especially lack of oxygen. When suitable conditions were restored, the spores give rise to bacilli again. When grown for several generations in pure cultures with no contact with an animal, these bacilli still caused anthrax. When Koch shared his discovery with Ferdinand Cohn, professor of botany at the University of Breslau, Cohn was astounded and, in 1876, had the work published in the botanical journal he edited. Koch became famous overnight in the world of science.

But in the contest between French and German scientists over the nature of immunity, Pasteur was a hero too. He sent back his honorary degree from Bonn University in protest against the war with Germany, while Koch's students constantly derided Pasteur's capabilities. In the impassioned debate about humoral versus cellular immune defenses, both sides came up with experimental evidence to support its own view. One theory bridging the two embattled sides came from the German scientist Paul Ehrlich. He proposed that antibodies recognizing the particular shapes of bacteria were naturally present on cells and released into the humors when they encountered the right germ. But the balance of new discoveries favored the humoral school, and in 1891 Koch proclaimed the demise of the cellular theory. The success of the humoral school was sealed by Behring and Kitasato's discovery of antibody and the clinical use of antibodies to fight diphtheria to such enormous effect in 1892.

It was by then the *fin de siècle,* an age of gaiety and decadence, sophistication and escapism. In the two decades following the siege of Paris in the Franco-Prussian War, the French capital had recovered and emerged as a vibrant center of artistic expression: the Symbolists were liberating poetry from past formality, capturing the immediate sensations of experience and hinting at the "dark and confused unity" of an inexpressible reality. But it was also the start of what was later called the *Belle Epoque*: the long period of peace in Europe which saw advances in technology that led to the automobile, the telephone, electric light, the phonograph, the cinematograph, and other symbols of the modern world. The coming of age of germ theory and the birth of immunology were very much a part of this era.

So how did antibodies, miraculous "weapons against death," recognize the vast universe of bacteria and other infectious agents? Chemists held the reins in immunological research and, based on their emerging understanding of how biomolecules were formed, came up with the suggestion that the shapes of bacteria became impressed on protein antibodies like keys fitting locks. These "instructive" theories of immunity held that the immune system learns the shapes of germs in order to fight them. In 1930 Felix Haurowitz and Friedrich Breinl, working in Prague, proposed that antibody was newly made on the surface of the germ, and this gave it the right shape. Ten years later the chemist Linus Pauling suggested more simply that the protein antibody folded into the right shape on the foreign structure. One year later the Australian Frank Macfarlane Burnet refined the theory and, taking account of the fact that these events might take place on cells, proposed that these cells then grew to produce numerous offspring, all making antibody of the same shape or recognition specificity.

An insurmountable problem for these instructive theories came with Watson and Crick's discovery, in 1953, of DNA: all information for protein shape is encoded in genes and so cannot be learned from subsequent encounters with microbes.

The alternative was hard to believe: to cover the possibilities, the immune system must pre-make all the shapes it might encounter in the universe of microbes. Neils Jerne, a Danish immunologist, suggested that all possible shapes (specificities) of antibody were indeed formed normally in the body, and the appropriate shapes were then selectively amplified when invading germs appeared.

Where and how this kind of selection and adaptation to infection took place was then tackled by Burnet, who proposed in the late 1950s that all shapes of preformed receptors are present on white cells, but that individual cells make only one shape, different from its neighbors. An invading germ fitting that particular profile triggers the rapid growth of that cell, whose many daughters then produced antibody all of the right shape. This "clonal selection" theory of immunity would stand the test of time.

But was antibody really the only frontline defender? The cellular school of thought was about to undergo a great renaissance. In the aftermath of the Second World War, when hospitals were struggling to cope with the dreadful burns suffered by combatants, skin grafting became an important, cutting-edge procedure. The British scientist Peter Medawar showed that skin grafts were frequently treated as invaders and rejected by a genetically controlled mechanism. He noted that rather than the usual inflammatory white cells, the inflammation of graft rejection contained mysterious cells called lymphocytes, a particular type of small white blood cell. In 1957 he began suggesting that cellular immunity mediated by lymphocytes might be important in tissue rejection rather than antibody. It was to be the dawn of a new age of cellular immunology.

We now know that one kind of lymphocyte (B lymphocytes) have the job of making antibodies. Another type, T-lymphocytes (it is the

T-cells that are gradually depleted in HIV patients), have the task of latching on to, and killing, virus-infected cells. T-cells also regulate immune responses. Given the hard-won knowledge of antibodies in the first half of the twentieth century (and leaving aside cellular immunity, which we will come to later), we can at last begin to imagine what happens in smallpox vaccination. The microbial shapes recognized by the immune system are small fragments of the whole germ, and each germ has a variety of such structures. But different germs occasionally possess at least some similar structures. These "cross-reactive" elements explain the relationship between smallpox and cowpox. When Jenner injected James Phipps with cowpox, the virus began to grow. The viral structures of cowpox that were similar to smallpox were recognized by the tiny minority of lymphocytes whose receptors matched these shapes. This triggered the rapid growth of these lymphocytes, and as their numbers grew they began to produce antibodies specific for both cowpox and smallpox. It was these cross-reactive, neutralizing antibodies that provided long-term protection against smallpox.

11

VICTORIOUS WEAPONS AGAINST ILLNESS AND DEATH

"Darkling I listen; and, for many a time
I have been half in love with easeful Death,
Call'd him soft names in many a mused rhyme,
To take into the air my quiet breath."

John Keats, "Ode to a Nightingale," 1819

Despite the lack of knowledge about how vaccines worked in the 1890s, three kinds of vaccine had been discovered. Edward Jenner had pioneered the use of a related (relatively harmless) virus against smallpox; Louis Pasteur had used weakened pathogens to protect against cholera and rabies; and he had also discovered a chemically altered pathogen to protect against anthrax. In addition, Emil von Behring had used an antibody (made in horses) to protect against diphtheria. All four were injected, but while the first three provoked a protective immune response, the fourth provided passive protection by directly disabling a deadly toxin, though the distinction wasn't clear at the time. The important thing was that each could make the difference between life and death.

In 1896 the *British Medical Journal* celebrated the centenary of Jenner's great discovery with a 50-page review of vaccination worldwide, but little else was done to mark the occasion. Things were different in Europe: in France and Germany there were public celebrations. In Russia the commemoration was postponed until October to coincide with the anniversary of Thomas Dimsdale's variolation of Catherine the Great.

Not that they still approved of variolation—Russia was the first to ban it in 1805. The Russians also reproduced all of Jenner's publications to mark his centenary year.

A prophet is never honored in his own country, and this was true for Jenner. Statues of him were erected in France, Italy, and Japan, but the British had a different way of doing things. Using private subscriptions from his friends, a statue had been installed in Gloucester Cathedral in 1825, but it would be another 30 years before a public statue was commissioned in England. In 1857 America, Russia, Italy, Switzerland, Norway, Prussia, and Holland all chipped in to the British fund. The statue, sculpted by William Calder Marshall, was duly placed in the southwest corner of Trafalgar Square, where Prince Albert unveiled it in 1858.

Trouble lay ahead. The statue was never really at home surrounded by the grandeur of what was then Britannia's Victory Square. Every other statue celebrated great military heroes and their deeds of conquest. Jenner wasn't even on a horse. It was also a time when the antivaccination movement was gaining in strength. Representations were made in Parliament for its removal, and the *Times* called for its relocation. After the death of Prince Albert, Jenner's steadfast advocate, in December 1861, the statue was quietly moved out. *Punch,* the satirical magazine whose cartoons had always followed the fortunes of vaccination, had this to say:

> *England's ingratitude still blots*
> *The escutcheon of the brave and free;*
> *I saved you many million spots*
> *And now you grudge one spot for me.*

The statue was moved to Kensington, where it stands to this day overlooking the Italian water gardens where one of London's lost rivers, the Westbourne, forever flows into the Royal Park. Here Jenner sits in reverie, his scrolled manuscript upon his knee. This is altogether a much better place for a man raised in the tranquil Vale of Severn and who was never at home in London. Two years after Jenner's uncelebrated centenary, the British Parliament passed the 1898 act recognizing conscientious objectors to vaccination and also banning arm-to-arm vaccination.

The latter measure was a little overdue. The new century quickly brought advances in manufacturing processes for medicines and related products. Antiseptic and aseptic techniques were applied in surgery, commercial production of sterile bandages began, advances in the manufacture of adhesives, foodstuffs, dyes, pigments, and other products were underway, and in 1909 the first synthetic polymer appeared in the form of Bakelite. By contrast, Jenner's vaccine remained unchanged.

Arm-to-arm vaccination had disappeared across Europe by 1900; instead, calves were used in vaccine production at increasing scale. Healthy animals were carefully selected and shown to be free of tuberculosis by a test based on the work of Robert Koch. The calves were quarantined in clean quarters, washed, shaved, and further cleaned with boiled water. "Seed vaccine" was injected into the prepared skin; six days later, when blisters developed, their contents—a clear fluid called lymph—were collected with sterile instruments. The donor animals were then killed and autopsied to make sure they were healthy. The lymph was homogenized with a grinder and mixed with glycerin, which preserves the vaccine in a fluid state and inhibits bacteria. In this way tens of thousands of doses in glass vials were produced. Inoculation consisted of placing a drop of vaccine on the skin and then pricking the area repeatedly with a needle to a depth that produced traces of blood. Early in the twentieth century, biochemists were already extracting and partially purifying proteins and other molecules from living tissues. The term *biochemistry* was coined in 1903, and in 1926 a protein was fully purified and crystallized by James Sumner, who showed that the enzyme urease is a pure protein. Other pure proteins soon followed. Vaccinology, however, continued on its own course. In the twentieth century, the material that eventually would be used throughout the vaccination campaign to rid the world of smallpox was a mixed soup of animal lymph from the blisters on calves and contained virus, skin scrapings, other cells, and debris.

Jenner's vaccine was a harmless relative of a deadly human virus, while Pasteur's vaccines were weakened or attenuated germs that naturally infect humans. Another early class made use of whole, killed bacteria. Vaccines against cholera, typhoid, and plague were produced like this at the end of the nineteenth century. This approach, in which bacteria or viruses are killed with chemicals such as formaldehyde, was used successfully for whooping cough from 1915 onward. In the course of the twentieth century, other kinds of vaccine were discovered. A logical extension of Behring's work on tetanus was to bypass the need for horse antibody by using a direct vaccine. This was eventually done by chemically treating deadly tetanus toxin to produce a harmless toxoid in work pioneered by the French veterinarian Gaston Ramon. Injected directly, the toxoid provokes protective antibodies.

The next major vaccine to be discovered was designed to protect against tuberculosis, or consumption as it was commonly called. But the development of a vaccine against this widespread infection had several very difficult false starts. Siddhartha Mukherjee, in his book *The Emperor of All Maladies,* characterizes cancer as paradigmatically belonging to the modern world, but he makes the points that tuberculosis (TB) was the emblematic disease of the nineteenth century, an illness so protracted that the fatal infection was defined not by death but by the long process of dying, often in tragic and romantic circumstances. This was true for Jenner's wife, Catherine, who died in 1815, and it was true for many important figures both in life and literature of the period. The illness, with its hectic fevers, melancholy torpors, and episodes of lucidity, was seen as a driving force for creativity and obsessive self-expression. Of course, TB was also widespread among ordinary working people whose means of livelihood depended in their physical strength, which was slowly but surely eroded by disease.

John Keats, a key figure in the second generation of Romantic poets, was often given to extremes of mood and he personifies the mercurial, brooding, consumptive temperament. Keats trained as a surgeon and apothecary and studied to become a doctor at Guy's Hospital, London, before devoting himself to poetry. He lost his mother to TB and selflessly nursed his brother Tom through his illness until Tom's death from consumption in 1818. Keats's poem "Ode to a Nightingale," written in 1819, expresses this experience of sickness and slow dying from which the poet, enraptured by the nightingale's song, longs to escape.

> Fade far away, dissolve, and quite forget
> What thou among the leaves hast never known,
> The weariness, the fever, and the fret
> Here, where men sit and hear each other groan;
> Where palsy shakes a few, sad, last gray hairs,
> Where youth grows pale, and spectre-thin, and dies . . .

His own consumptive illness became evident in 1820. Keats suffered two lung hemorrhages early in the year, losing large amount of blood; the treatment for his condition prescribed by the medical profession was, of course, therapeutic bleeding.

Tuberculosis is a progressive chronic disease caused by *Mycobacterium tuberculosis,* as Robert Koch discovered in 1882 by using groundbreaking staining and culture techniques. Koch named the offender the *tubercle bacillus.* Principally a lung infection, TB can spread to other organs. The infection is transmitted when someone with active TB coughs, projecting an aerosol of infectious pathogens. Although a single bacillus may cause infection, the great majority of individuals exposed to these mycobacteria (over 90 percent) do not develop tuberculosis. The invading bacteria are either killed and eliminated there and then, or else their growth is limited by the immune system. In the susceptible minority, a protracted battle between the bacillus and the immune system ensues. Foci of bacteria in the lungs are trapped through a chronic walling-in process in which the infection is contained by a granuloma. This is an enclosing ball of tightly packed immune cells called macrophages (from the Greek, *bit eaters*). Usually wandering cells, the macrophages in a granuloma become more like structural skin cells and may exist in a fibrous framework. In this contained state the bacteria may remain dormant for years or even decades. In a process that is poorly understood, the latent *M. tuberculosis* can reactivate and cause disease. (Any impairment of immune function can result in reactivation.) The patient with active TB is afflicted by coughing, chest pains, weakness, fatigue, anorexia, weight loss, fever, and night sweats.

Until the 1940s there was no effective treatment for TB, and Keats's experience is not untypical of how the illness progressed. During 1820 Keats displayed increasingly serious symptoms of tuberculosis, suffering lung hemorrhages and losing large amounts of blood. The treatment meted out by the attending physician was yet more bleeding. At the suggestion of his doctors, Keats agreed to move to Italy with his friend, the painter Joseph Severn. He left England in September, and after a voyage of storms and calms they at last docked in Naples, where the ship was quarantined because of a suspected outbreak of cholera in Britain. Keats reached Rome in mid-November and took lodging in a villa overlooking the Spanish Steps. Despite care from Severn and an English doctor practicing in Rome, Dr. James Clark, his health rapidly deteriorated. The treatments probably hastened his death. Clark initially declared the illness due to "mental exertion" and said the disorder centered on the stomach. Eventually he diagnosed consumption and placed Keats on a starvation diet of an anchovy and a piece of bread a day, hoping to reduce the blood flow to his stomach, and applied repeated bleeds that left the poet dizzy and confused. One day in mid-December the bedridden Keats, coughing and vomiting blood, swore that the day would be his last. His friend Severn, fearing a suicide attempt, withheld the laudanum prescribed by Clark to alleviate Keats's suffering. Keats was delirious for the rest of the day until a violent hemorrhage and bleeding weakened him into calm. Over the next nine days he suffered severe hemorrhages and repeated bleedings by Clark and he asked his doctor how long this "posthumous life" would drag on.

It dragged on another two months. His death was probably typical of how tuberculosis sometimes ended. The poet was frequently delirious with fever, and Severn reported that he often cried on waking at finding himself still alive. On the day of Keats's death, on February 23, 1821, Severn wrote: "Keats raves till I am in a complete tremble for him." Then came lucidity, in which Keats told Severn not to be frightened, because he would now die easily, and thanked God that it had come at last. He survived another seven hours with phlegm "boiling in his throat" until he sank into death. This was the disease for which the great Robert Koch promised a cure in 1890.

Robert Koch's identification, in 1882, of the pathogen that causes TB was hailed as a "world-shaking" event and made him famous far beyond the scientific establishment. It was generally believed that this must signal the imminent defeat of this terrible disease. Eight years later, at the Tenth International Medical Congress in Berlin in August 1890, Koch made the sensational announcement that his studies had produced a therapeutic remedy that could halt the progress of tuberculosis. When this treatment became available in October, it received a euphoric reception. Koch claimed that when guinea pigs were experimentally infected with tuberculosis, the pathological process of the disease could be brought to a complete standstill by his therapeutic agent. He did not disclose its nature or identity. Throughout the winter of 1890, the effects of the miracle remedy were studied in patients with TB, but as the new year dawned, it was evident that all was not well. The secret medicine had failed to demonstrate any beneficial therapeutic effects. This put a great deal of pressure on Koch, and early in 1891 he at last revealed what his remedy was. It was a glycerin extract of tubercle bacilli, which he called *tuberculin*. This is the kind of killed, fragmented preparation of bacteria that might constitute a vaccine, but ideas about provoking an immune response had yet to be developed. It appears Koch's theories had more to do with starving the TB bacteria by depleting elements essential for their survival. Koch interpreted the dying away of tissue (necrosis) around sites of bacterial infection in guinea pigs given tuberculin as an indicator of success, but in fact this kind of tissue death is not an indication of arresting the disease.

Toward the end of the year, Paul Baumgarten, who had described the tuberculosis bacillus independently of Koch in 1882, gave a devastating summary of the animal testing of tuberculin: large doses caused damage in guinea pigs with TB, and small doses gave no benefit at all. Simultaneously, reports of deterioration among patients undergoing treatment with tuberculin began to appear. Tuberculin was finished as far as the medical establishment and the public were concerned. There is no indication that Koch was deliberately misleading about the supposed effect of tuberculin. He seems to have firmly believed that tuberculin was curative, based on the principle of starving the bacterium. He kept working on the issue, presenting an improved tuberculin in 1897; he remained faithful to his remedy until at least 1901, when he finally gave up on it. In the interim, the value of tuberculin as a diagnostic agent began to be appreciated. Today, a version of tuberculin (sometimes called purified protein derivative) is used as an indicator of TB infection, both current or at some time in the past. The superficial skin test measures the specific immune response to TB proteins, which is evident as a raising of the skin surface after 48 hours.

During the later years of his life, Koch came to the conclusion that the bacilli that caused human and bovine tuberculosis were not identical. His presentation of this at the International Medical Congress on Tuberculosis in London in 1901 caused great controversy and opposition. We now know that Koch was right. In modern nomenclature the human pathogen is termed *Mycobacterium tuberculosis,* while the bovine germ is called *Mycobacterium bovis.* As early as 1854, Jean Antoine Villemin had realized that cattle, as well as humans, suffer from TB, The fact that Koch was able to distinguish between the two germs posed a new question: could cow TB bacilli do for human tuberculosis what cowpox had done for smallpox? Tragically, Italian clinical trials attempting to use the bovine bacillus in this way ended in disaster—the bovine germ was just as dangerous in humans.

In 1908, French bacteriologist Albert Calmette and his assistant Camille Guérin, a veterinarian, working at the Institut Pasteur in Lille, isolated bacteria from a cow with TB. They carefully cultured the isolate, transferring it every three weeks into fresh culture medium. This practice, called passaging (pronounced pas*arg*ing), was maintained throughout World War I until 1919. The preparation, which had once formed rough and granular colonies, slowly changed until the colonies were round and smooth. During its long artificial passage, many undefined genetic changes had occurred. When the strain was retested in cows after thirteen years, it had dramatically lost its virulence and no longer produced disease. This would mean little unless the bacilli had retained enough of the original structure to provoke the immune responses that protects against human TB.

The BCG (Bacillus Calmette Guérin) vaccine was first tested in humans in 1921 and did indeed induce the right kind of immunity. Nearly a half-million infants were safely vaccinated. Then, in the summer of 1930 in the Hanseatic city of Lübeck in northern Germany, disaster struck. German authorities were becoming increasingly interested in the possibility of protecting children against TB using the French bacillus, and trials in Dusseldorf and Berlin were going well. The health authority in Lübeck sought to follow suit and purchased BCG culture, which was sent in bulk from Paris. The newborn children of the city, once their parents had agreed, were duly immunized. Two hundred and forty babies were vaccinated in the first ten days of life.

Almost all the babies developed acute tuberculosis. Seventy-two of the infants died. The *Lancet,* the internationally respected medical journal, carried breaking news of the tragedy in May 1930, when fourteen newborn babies had already died and 50 were severely ill. The article reported how the vaccine had been sent from Paris and then divided and subcultured in a local laboratory. No one knew what was going so terribly wrong. Were the unsupervised midwives making mistakes? Was BCG really safe, or could the fears of rogue colonies reverting to the deadly original bacillus be true after all? The distress and despair of grieving parents and the dismay and confusion of health professionals is almost impossible to imagine. Experts called into question the wisdom of live virus vaccines, pointing to the new plague and typhoid vaccines that used safely killed germs. "Who knows," one expert said, "how long an attenuated bacillus can lie dormant and then assume its former virulence?"

It was subsequently discovered that the Lübeck BCG had been contaminated with a virulent bacillus grown in the same incubator. A technician received a two-year prison sentence. BCG is now used around the world, although it remains controversial and has never been employed for mass vaccination in the United States, where reliance is placed instead on the detection and treatment of latent tuberculosis. One drawback of mass vaccination with BCG is that all recipients become positive in the tuberculin skin test, meaning it is no longer possible to screen populations for latent TB. At best, the vaccine is 80 percent effective in preventing tuberculosis for a period of around fifteen years, but BCG is much less effective in tropical regions for reasons not yet understood. The world is very much in need of a better vaccine against TB. As to the wisdom of live vaccines, today we have nine safe and highly effective live attenuated vaccines in routine use against bacterial and viral diseases, including measles, mumps, rubella (German measles), typhoid fever, influenza, and rotavirus, which causes infant enteritis. The latter saves many lives because diarrhea caused by rotavirus kills more than a million children worldwide every year.

In the early 1960s, at age thirteen, I took part in a large-scale assessment of BCG in British schoolchildren (at the time I didn't realize it was a vaccine study). First, we all submitted to the tuberculin skin test that arose from Koch's work, which was applied with a clever multineedle gun (Heaf gun). The BCG injection was quite painful, so I was relieved when my skin became raised and red—a very positive result that meant I wouldn't be given the vaccine. But of course, I soon reflected on what this might mean. I remember taking the bus to Cambridge for an X-ray to find out if I was doomed.

I grew up on a fenland farm. My mother, a farmer's daughter, was also a milkmaid, and I loved to drink raw milk, warm from the cow. Neither the Heaf test nor the Mantoux test, which replaced it in the United Kingdom, discriminates between human and bovine TB, which can also infect humans. Perhaps I was exposed in this way. But I also remember another occasion. I was riding on a country road one summer's day with my friend from school named Chris. The fenland landscape is a flat one, especially to a small child who sees his world diminishing in all directions, each layer of distance draped in a softer blue-gray veil. In the center of this empty vastness the two of us rode our bicycles side by side. After a while I ventured to tell Chris a joke. I can't remember what it was, but it certainly made him laugh. Helpless with laughter, he protested he'd laugh until he coughed up blood, "You know," he said, "when you laugh too much and start to cough, and then you cough up blood."

Of course, I didn't know, and told him so. I told him it had never happened to me. I said it wasn't right, and that I was worried about him. I never found out what happened to Chris: soon afterward his family moved to another part of the country. Perhaps this friendship explains my positive Heaf test six years later.

I got off the bus in Cambridge and dragged myself up Castle Hill, the only hill in the county. Somewhere in the bowels of Shire Hall, I found the clinic. Feeling weak and fearful, I waited to be summoned. The examining physician, who peered for an eternity at my X-ray plate, was quite devoid of sympathy. But that didn't matter when he finally broke the news: my chest X-ray was clear; I was free to go. I left the building and ran up Castle Mound (the hill on top of Castle Hill), exulting at the future which now stretched out below me into a vanishing distance.

Until September 2005 all U.K. children between ten and fourteen were offered BCG vaccination. But because of much lower TB risk these days, the policy has changed. The vaccine is now reserved for children at elevated risk, often because of where they live, which is sometimes finely judged across different regions in the same city. My grandson Stanley, who was born near Hackney in East London, was given BCG vaccination as an infant, but his baby brother Albert, born ten miles farther north in Muswell Hill, was not.

At the beginning of the twentieth century, the sanatorium movement, which espoused the therapeutic virtues of fresh air, bed rest, and sunlight, reached its peak in Europe and the United States. Secluded sanatoriums were often built in beautiful locations among lakes and mountains or overlooking the sea. Thomas Mann, in his novel *The Magic Mountain,* captures the symbolic significance of an ascent from the mundane realities of everyday life into the rarefied seclusion of an Alpine TB sanatorium. Less privileged sufferers had to make do with outdoor beds in grim infirmaries at the margins of industrial towns. In 1944, 54 years after Koch's announcement of a remedy, the first truly effective treatment for TB, the antibiotic streptomycin, was developed—but the problem of resistance soon emerged. In 1952 another key drug, isonazid, was discovered, and it remains the cornerstone of therapy today. Drug combinations were optimized in the 1960s to produce a daily-dosing curative treatment two years long. In 1972, adding a new drug, rifampicin, reduced daily dosing to nine months. More drugs followed, and today a six-month combination regimen can cure TB.

In the middle of the twentieth century, TB was reduced dramatically, but in the 1980s, the combination of HIV, emerging drug resistance, and socioeconomic factors permitted the return of TB in high-risk groups in the cities of the developed world. Today TB in richer countries is a disease of poverty and homelessness, afflicting the poor and those made vulnerable by deprivation, drug addiction, and alcohol abuse. But by far the greatest burden of TB falls on the developing world, where 95 percent of cases are found and where BCG vaccination is often ineffective. About a third of HIV victims in sub-Saharan Africa die of tuberculosis. Around one third of the world population has TB, though for many it exists as a latent infection controlled by the immune system. The World Health Organization (WHO) estimates that in 2010, just under 9 million people fell ill with TB, and 1.4 million died. But the picture is once more improving: the estimated number of people falling ill with tuberculosis each year is declining, although very slowly, which means that the world is just about on track to achieve the WHO Millennium Development Goal to reverse the spread of TB by 2015.

Although BCG, the most widely used vaccine in history, fails to prevent primary infection in much of the developing world, the vaccine is still used because it reduces dissemination of TB through the body, which in turn lowers newborn and childhood death rates. But a better vaccine is sorely needed. Several new vaccines are being tested, and one interesting newcomer exploits a new concept. To tackle the thorny problem of targeting dormant TB as well as active bacteria, researchers in Pittsburgh and Copenhagen have put together several purified bacterial structures that are recognized by immune cells. Crucially, these building blocks appear at various stages in the bacterial life cycle, including structures appearing in response to stress when bacteria are starved of nutrients or oxygen. The aim of this multistage vaccine is to control both active *and* latent bacteria to overcome the great problem of reactivation. On their own, these purified subunits would not convince the immune system that a dangerous invader was on board, and so would not induce a protective immune response. But the new vaccine includes a hefty *danger signal* in the form of microbial signature molecules that trick the first-line defenses of the immune system into sensing a dangerous invader in the body. The new vaccine, whose experimental name is H56, entered the first stages of clinical trials in Cape Town in December 2011. There are ten other new vaccines for TB in the early stages of testing. There is no shortage of cautious optimism, but despite immunizing more than 3 billion people with BCG, a mystery still surrounds the shifting relationship between the immune system and the TB bacterium.

12

FIRST LIGHT ON THE MYSTERY OF INFANTILE PARALYSIS

"The light in this room is of a lamp. Its flame in the glass is of the dry, silent and famished delicateness of the latest lateness of the night, and of such ultimate, such holiness of silence and peace that all on earth and within extremist remembrance seems suspended upon it in perfection as upon reflective water: and I feel that if I can by utter quietness succeed in not disturbing this silence, in not so much as touching this plain of water, I can tell you anything within realm of God, whatsoever it may be, that I wish to tell you, and that what so ever it may be, you will not be able to help but understand it."

James Agee, *Let Us Now Praise Famous Men*, 1936

The tragedy of tuberculosis lay in the lingering decline of those infected. In the nineteenth century, it eluded every remedy. The episodes of remission merely served to kindle false hope before the illness returned to pursue its life-consuming course. But another hugely feared disease was terrible for a different reason: the cruel suddenness with which it struck its victims down, apparently at random, transforming the healthiest, most energetic children into lifelong invalids in the span of a day and a night.

The story of this illness, and the battle to defeat it, unfolded not in Europe but in the United States in the first half of the twentieth century. In the summer of 1894, an epidemic of paralysis descended on a small town in New England. It happened in Rutland, a marble-mining community in the Otter Valley, which is set deep amid the lakes and forests of Vermont. The epidemic might have gone unnoticed but for a young country doctor, Charles Caverly, with a passion for public health. He carefully traced all 123 cases, recording that 50 remained permanently paralyzed while eighteen victims died. The great majority were boys of less than six. Caverly thought playing too hard in the hot sun might increase susceptibility to the paralytic condition, a notion that would stick for the next 50 years. Caverly realized that the contagion could also manifest itself as just a minor sickness with no lasting ill effects.

The disease that Caverly described was infantile paralysis, or poliomyelitis, commonly known as polio. In the first decade of the twentieth century, polio reached as far as New York City. In 1907, 2,000 cases were reported, and similar outbreaks were soon occurring across the United States. Between 1910 and 1914, Massachusetts, Minnesota, Nebraska, Ohio, and Wisconsin were affected. By this time the first great step in understanding the disease had been taken. In 1908, the Austrian physician Karl Landsteiner, already distinguished for his discovery of blood groups, identified the cause of polio, which was also afflicting Europe. Together with his colleague Erwin Popper, he made an emulsion of nerve tissue from a boy who had died of polio and succeeded in transferring the paralytic disease to laboratory monkeys. Crucially, he found the infectious agent could pass through porcelain filters and must therefore be a virus.

From quite early on, public health records seemed to show that, rather than being associated with poor sanitation and underprivileged communities, polio hit the more advantaged neighborhoods. This became significant as the nature of polio began to emerge. In 1916 a major outbreak struck America, and 27,000 people suffered paralysis. Six thousand died. New York City was badly hit once more, and quarantine was one of the few weapons available against the epidemic. The New York outbreak began in June in Pigtown, a small community of Italian immigrants in Brooklyn; it soon spread to other boroughs. The worst hit seemed to be the privileged communities of Staten Island, where sanitation standards were high. By July all children leaving the city were required by the health department to have a certificate confirming they were free of polio, while police in the surrounding towns turned back desperate city-dwellers fleeing the epidemic. By August, despite these emergency measures, the polio outbreak had spread to New Jersey, Connecticut, Pennsylvania, and upstate New York.

We now know that severe polio is rare in communities with poor hygiene, probably because babies are exposed to the virus while still protected by antibodies in a mother's milk. In this situation, most of the community has natural immunity, and infection produces, at worst, a mild transient sickness. Improvements in waste disposal and domestic sanitation in the early twentieth century may partly explain the epidemics of paralytic polio, which began to occur in the developed world, largely in cities during the summertime. After 1894, polio epidemics broke out almost every summer in the United States, a tragic cycle that would continue to the 1950s and cast a shadow of fear over the vacation seasons. The American obsession with the virtues of cleanliness, enthusiastically promoted by soap and disinfectant manufacturers, didn't help the problem and may have made it worse by reducing early exposure in infants protected by maternal antibodies. By the time of the Great Depression, paralytic poliomyelitis was a hugely feared disease.

Among the foremost authorities on polio in the United States was the director of the Rockefeller Institute for Medical Research in New York City, Simon Flexner. Already distinguished for his work on bacterial meningitis, Flexner replicated the observations of Landsteiner in Vienna and went a step further: he transferred polio between laboratory monkeys, establishing an animal model for the disease. Determined to discover how polio invades the body, Flexner compared the effects of administering viral material through the mouth and nose and found that only monkeys infected through the nose went down with the infection. He concluded that polio was a disease of the nervous system, passing directly through nasal nerves to brain and spinal cord. This led him to abandon his ideas for a vaccine, since the neural tissue would be inaccessible to protective antibodies in the blood. Instead he proposed measures to prevent infection by nasal blockade.

Flexner's ideas dominated thinking on polio prevention throughout the 1920s and 1930s. As Director of the lavishly resourced Rockefeller Institute, he fostered research on nasal blockade, and in 1936 he encouraged his head of virology, Peter Olinsky, and his young associate, Albert Sabin, to work on the problem. Attempts to prevent the spread of polio with chemical nasal sprays were mounted in the field by public health officials desperate to limit summer epidemics. Many children were made to endure the treatments with no indication of success.

In the summer of 1921, polio, or infantile paralysis, was still a relatively new disease notorious for afflicting children. It was therefore a shock to many Americans when Franklin Delano Roosevelt, the Democratic contender for the vice presidency who was famous for his energy and robust constitution, was struck down by the illness at the age of 39. Roosevelt had been summoned to the capital in the simmering heat of summer to answer questions on a scandal in the Navy Department, where he was assistant secretary. The investigation did not go well for Roosevelt. In need of respite he headed for Campobello Island, his family's holiday retreat off New Brunswick, north of Maine, stopping on the way to attend a Boy Scout jamboree where he marched among the ranks of youngsters. August 8, the day after his arrival, was filled with energetic sailing, swimming, and arduous running with his own tireless brood. By evening he was feeling strange, with aches and chills he'd never felt before. The next morning his left leg was all but immovable. His illness rapidly worsened and he quickly lost the use of both legs. Polio is characterized by flaccid, or floppy, paralysis. The patient is unable to tense the muscles of the useless limbs, which remain relaxed and eventually shrink and waste away. Doctors were called, including a specialist on infantile paralysis summoned from Boston who diagnosed poliomyelitis.

The return of the Roosevelts to New York marked the beginning of a lifelong campaign to prevent the public from knowing of the great man's permanent physical impediment. The press saw only a beaming FDR framed in the windows of trains and automobiles, and the public was assured that he would have no lasting disability.

Roosevelt returned to work in the fall of 1922. By this time his upper body was extremely strong from the exercise regimen he kept up to compensate for his paralyzed legs. He now wore steel and leather braces from hip to ankle. FDR threw himself back into the pursuit of his career. He set up a new law practice and appointed a dynamic young lawyer, Basil O'Connor, to be his partner in 1924. Leaving his young partner to get on with running the business, Roosevelt sought out the benefits of hydrotherapy at a mineral springs resort in Meriwether County, Georgia, owned by the power and railroad magnate George Foster Peabody. For Roosevelt, the somewhat run-down Meriwether Inn in the backwoods of Georgia seemed the perfect retreat in which to hide from public view and begin to recuperate. The warm waters of the natural springs, made buoyant by high mineral content, were ideal for lifting and soothing flaccid limbs. Roosevelt was joined by his wife, Eleanor, who was shocked by the primitive conditions, the constitutional racism, and the day-to-day struggles of the southern poor. She didn't tarry long.

Inevitably the press took an interest in Roosevelt's sojourns at Warm Springs. It reported enthusiastically on the medicinal properties of its waters tucked away in the backcountry and painted a picture of a man who would soon be restored to health. When Roosevelt returned to Warm Springs a few months later, he found a change had taken place: alongside the usual customers, a number of polio victims had made the pilgrimage, and more were on their way. Calling themselves *polios* in solidarity, the new visitors upset the regular guests. Some feared the paralysis might be contagious; others treated the polios as objects of curiosity to be stared at from a safe distance. Roosevelt soon took things in hand, organizing a degree of separation for the convalescents, who were given their own dining room and bathing pool. From here his vision for Warm Springs—as a haven where the disabled could retreat, receive therapy, and work toward rehabilitation—began to be realized, and he decided to purchase the resort. Eleanor Roosevelt firmly believed that had her husband not been elected governor of New York in 1928, he would have stayed on as director of his Institute for Rehabilitation.

When Roosevelt became president in a landslide victory in 1932, Americans saw him as the energetic, optimistic leader who would lead them out of the Great Depression. The general public had been shielded from the extent of his physical handicap, but many of the victims of polio also saw in him the supreme role model of a man transcending disability. As he assumed office, he left his Warm Springs Foundation in the care of Basil O'Connor, his young partner. Faced with the task of financing a charity in the worst economic downturn in modern history, O'Connor, whose working-class Irish roots hadn't stopped him from earning a place at Harvard, set about hiring resourceful managers and fund-raisers. Somehow they succeeded in carrying the optimism of the president's New Deal into the realm of philanthropy with a series of ingenious FDR-linked fund-raising initiatives. In particular, FDR Birthday Balls were held annually in numerous cities across the country, which raised funds to keep the Warm Springs Foundation afloat through the hardest of times. In 1938 Roosevelt announced the formation of a National Foundation for Infantile Paralysis, whose aim was to find the cure for polio and provide the best treatment. He appointed Basil O'Connor as director.

The National Foundation would become the supreme model for private charitable fund-raising, quickly growing into the largest voluntary health organization in America. Its success in promoting awareness, mobilizing communities, sowing the seeds of hope, and collecting funds for the care of patients and the search for a cure would be unparalleled in the twentieth century. One of its most successful strategies was harnessing the power of show business. Its public relations operation quickly developed units for both radio and motion pictures, enlisting the help of movie stars and entertainers much loved by the American public. The first of these was Eddie Cantor, who came up with the "March of Dimes" slogan, adapted from the newsreel title "March of Time," which was popular in all movie theatres.

The March of Dimes would come to symbolize the power of individual generosity at the grassroots level across the United States. The breadth of Cantor's popularity was based less on his songs and stage performances and more on his radio broadcasts about home life with his wife, Ida, and their five daughters. As with Roosevelt, the power of the intimate, disembodied voice, speaking in the shadows at the fireside in millions of households across the country, made Cantor a friend of the family for ordinary people. In 1938, the great majority of country homes had no electricity, and television would not arrive until the 1940s. But battery radio reached everywhere, beyond the glow of cities to the oil-lit living rooms of rural America. In 1938 Cantor launched the March of Dimes on radio, and the broadcast was followed by appeals from Jack Benny, Bing Crosby, the Lone Ranger, and other household names.

Even before the March of Dimes appeal, a proportion of the money raised for the Warm Springs Foundation was used to fund research. Grants were made, among others, to New York University School of Medicine. A young Canadian researcher, Maurice Brodie, was hired to work on a potential vaccine against polio. Brodie sought to exploit the principles already pioneered successfully for tetanus, diphtheria, cholera, and plague, in which the vaccine material is killed by chemical cross-linking with formaldehyde. He used poliovirus grown in monkey spinal cord (the only laboratory source of the virus). Supported by his department head, William H. Park, a long-time rival of Simon Flexner, Brodie inoculated monkeys, demonstrating that the animals produced antibodies to the injected virus. Park and Brodie then tried the vaccine on themselves, experiencing only minor discomfort at the injection site. A dozen children, with the consent of their parents, were then vaccinated. Park and Brodie published their results in the *Journal of the American Medical Association* in 1935, concluding that the vaccine was probably safe for use in humans.

It is perhaps a measure of the fear of polio, and the intense pressure on public health authorities to act, that officials in some states permitted trials to go ahead. The results were inconclusive. The trials were conducted haphazardly and provided no evidence that the vaccine protected against polio. Because the vaccine was produced in monkey neural tissue, it carried the danger of provoking allergic responses to a monkey neural protein (myelin basic protein), which in turn could result in immune attack on normal, healthy nerve tissue. When larger numbers of children were injected, the Brodie Park vaccine did not induce a sufficient level of protective antibodies and occasionally produced severe allergic reactions. Moreover, researchers at the Rockefeller Institute were unable to reproduce the earlier results published by Park and Brodie. Another researcher, Philadelphia-based pathologist John A. Kolmer, produced a rival vaccine in the same year. His used a chemically attenuated living virus to create a harmless immunizing infection. Funded independently of Roosevelt's foundation, Kolmer conducted experiments on monkeys and proceeded to vaccinate himself, his two sons, and 23 other children. Encouraged by the absence of ill effects, he went on to vaccinate 10,000 children. But the trial of Kolmer's vaccine produced alarming results: nine children died of polio, their deaths attributed to the vaccine.

With the benefit of hindsight, it's clear that much more painstaking work was required to understand the right conditions in which to inactivate poliovirus and how best to measure any residual biological activity. These first attempts were made too early and in haste; both Kolmer and Brodie had to come to terms with the harm these vaccines caused. Brodie bore the greater burden. His career was irretrievably blighted and he died prematurely at the age of 36 amid rumors of suicide. At the time these early vaccines were tested, the existence of different types of poliovirus was not appreciated. Neither were the dangers of unknown monkey viruses that could contaminate vaccines grown in monkey nerve tissue. But at the great cost of tragedy for the families affected, the two failed vaccines at least set the scene for what would happen later and spurred fundamental research on the biology of poliovirus.

Yet even with a safe vaccine and properly conducted trials, wouldn't a vaccine prove futile because of the poliovirus's path of entry to the nervous system? The eminent investigator Simon Flexner had established a direct neural pathway in his monkey studies, but the brain and peripheral neural tissue are not readily accessible to the immune system.

The answer to this question was surprising. It gradually became evident that Flexner was entirely wrong about the route of entry of the infection. He had been led astray by his exclusive reliance on his monkey model of polio. The species he chose to study—*Macaca mulatta,* or rhesus monkey—is not susceptible to polio by the natural route of infection. The virus doesn't replicate in the rhesus monkey gastrointestinal tract, and these monkeys can *only* be infected through the nasal nerves or by direct injection into the brain. In humans, the only natural host to the virus, the situation is quite different: polio is a gastrointestinal virus that replicates in the cells lining the gut and readily passes between individuals through fecal contamination of food and water. Only in 1 percent or less of natural cases does the virus enter the nervous system, where it damages nerves controlling muscles (motor nerves) and produces the characteristic flaccid paralysis.

Flexner is remembered for mentoring a generation of influential leaders in polio research, but not for discovering the route of entry or the primary site of poliovirus growth. The respect he commanded, and the influence he wielded, was such that many investigators were misled. Because he transferred, or *passaged,* the virus many times through monkey brains and spinal cords in the laboratory, he selectively favored the strains that were best at growing in nerve tissue and eliminated those that grew poorly in nerves. The strain that eventually emerged, called MV virus (or mixed virus), was exclusively neurotropic (restricted to nerve tissue). Since Flexner supplied this strain to numerous research laboratories, many investigators were misled by this atypical virus. As the legions of sufferers accumulated year by year, the need for a vaccine grew and grew. But after the failed vaccine attempts of 1935 and the lack of knowledge about poliovirus biology, it would be twenty years before polio vaccine researchers were ready to try again.

Plate 1. The plague doctor of the 17th and 18th century wore a broad brimmed hat and a large, beak-like mask filled with straw, balm-mint, camphor, cloves, laudanum, myrrh, rose petals and other aromatic principles against the foul miasmas believed to cause disease. His eyes were protected by colored glass, and his full-length waxed leather or canvas coat was tightly laced and buttoned so that noxious agents could not penetrate. His duties to the civic authorities who hired him included care of plague victims, as well as reporting and recording deaths. Oil painting by Leonie Rhodes, 2013, photographed by Daniel Michaud. Reproduced by kind permission of the artist.

Plate 2. Lady Mary Wortley Montagu in Turkish Dress. *Reproduced by kind permission of the Wellcome Library, London.*

Plate 3. Edward
Jenner *by John
Raphael Smith, 1800.
Reproduced by kind
permission of the
Wellcome Library,
London.*

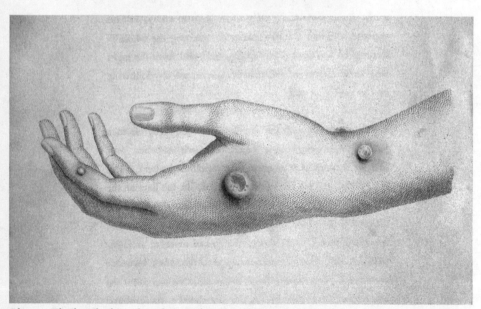

Plate 4. The hand of Sarah Nelmes infected with the cowpox. From An inquiry into the
causes and effects of the variolae vaccinae. *By Edward Jenner, 1798. Reproduced by kind
permission of the Wellcome Library, London.*

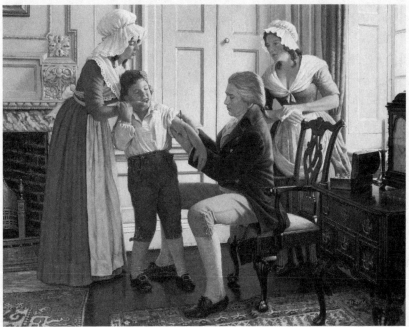

Plate 5. Jenner: Smallpox is stemmed *by Robert Thom, circa 1960.*
Reproduced with permission from the Collection of the University of Michigan
Health System, Gift of Pfizer, Inc., UMHS.23.

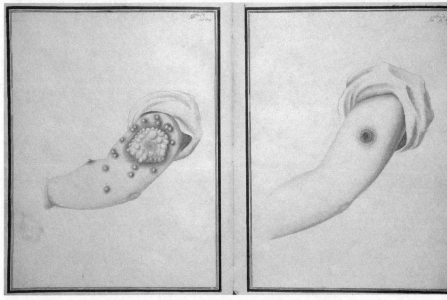

Plate 6. A print comparing smallpox inoculation (variolation) and cowpox inoculation
(vaccination) after 14 days from 1802. Reproduced by kind permission of the Wellcome
Library, London.

Plate 7. The Cow-Pock—or—the Wonderful Effects of the New Inoculation!, 1802. James Gillray portrays Edward Jenner among patients in the Smallpox and Inoculation Hospital at St. Pancras in this cartoon ridiculing the exaggerated fears of those opposing vaccination. Reproduced by kind permission of the Wellcome Library, London.

Plate 8. The inaugural meeting of the Medical Society of London in the Society's Council Chamber. Samuel Medley, 1800. John Coakley Lettsome, depicted addressing the learned assembly, ensured that Jenner's image (back row, 5th from left) was added retrospectively. Reproduced by kind permission of the Medical Society of London and the Wellcome Library, London.

Plate 9. Vaccination against Small Pox or Mercenary and Merciless spreaders of Death and Devastation driven out of Society, *Isaac Cruikshank, 1808. Jenner and two colleagues are shown driving away three variolaters. Infants, dead of smallpox, lie scattered at their feet. Reproduced by kind permission of the Wellcome Library, London.*

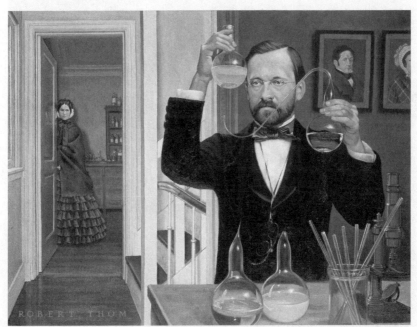

Plate 10. Pasteur: The Chemist Who Transformed Medicine *by Robert Thom, circa 1960. Reproduced with permission from the Collection of the University of Michigan Health System, Gift of Pfizer, Inc., UMHS.23.*

Plate 11. Portrait of Robert Herman Koch (1843–1910), the founder of modern bacteriology, 1890. Reproduced by kind permission of the Wellcome Library, London.

Plate 12. Sir Frank Macfarlane Burnet developed the theory of adaptive immunity which explains how the immune system responds to infection and protects against subsequent attacks by the same infectious agent. Reproduced by kind permission of the Wellcome Library, London.

Plate 13. President Franklin Delano Roosevelt with Fala the dog and Ruthie Bie, a friend's granddaughter, at Hill Top Cottage, Hyde Park, New York. Courtesy of the Franklin D. Roosevelt Presidential Library.

Plate 14. *Albert Sabin, Jonas Salk, and Basil O'Connor. Salk developed a killed virus polio vaccine, Sabin developed a live virus vacccine and O'Connor led the March of Dimes campaign which made this possible. Reproduced by kind permission of the March of Dimes Foundation .*

Plate 15. Christina's World *by Andrew Wyeth, 1948. Wyeth produced many studies of his neighbor in Maine, Christina Olsen, whose paralysis is generally attributed to polio. These made her perhaps the most famous model in modern American art.* Christina's World *is the most iconic of these works and it was also her personal favorite. © Andrew Wyeth, 1948, tempera, 32¼ x 47¾. Digital Image © (2013), Museum of Modern Art, New York/Scala, Florence. Reproduced by kind permisssion of the Wyeth Foundation*

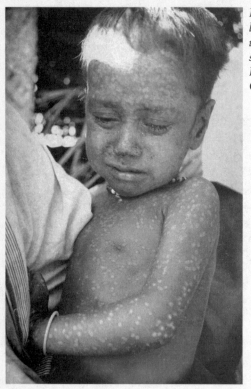

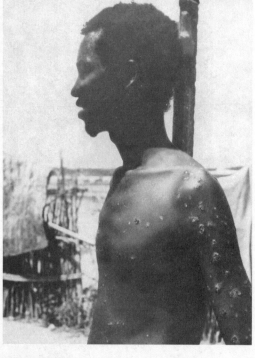

Plate 16. Rahima Banu, at age 3, was the last case of naturally occurring infection with Variola major, *the severe form of smallpox. She recovered from her illness. Photograph by Stanley Foster, courtesy of CDC and the World Health Organization.*

Plate 17. Ali Maow Maalin, at age 23, was the last person to be naturally infected with smallpox. Photographed here in 1977. Courtesy of J. Wickett, Center for Disease Control and the World Health Organization.

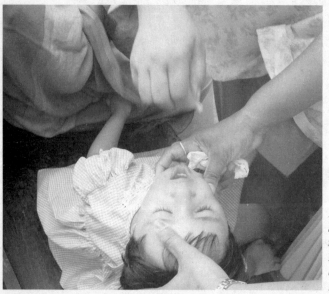

Plate 18. Young girl being given a Sabin oral polio vaccine as part of the National Immunization Day in Bangladesh in 2006. Reproduced by kind permission of Rafiqul Islam, Wellcome Images, London.

13

YEARNING TO BREATHE FREE

"She was shaking from head to toe. She complained of a painful stiffness in the upper vertebrae—and I thought of poliomyelitis as any American parent would."

Vladimir Nabokov, *Lolita*, 1958

Escaping the city's heat entirely and being sent off to a summer camp in the mountains or the countryside was considered a child's best protection against catching polio . . . So the privileged lucky ones disappeared from the city for the summer while the rest of us remained behind to do exactly what we shouldn't, given that "overexertion" was suspected of being yet another possible cause of polio: we played inning after inning and game after game of softball on the baking asphalt of the school playground, running around all day in the extreme heat, drinking thirstily from the forbidden waterfountain, between innings seated on a bench crushed up against one another, clutching in our laps the well-worn, grimy mitts we used out in the field to mop sweat off our foreheads and to keep it from running into our eyes—clowning and carrying on in our soaking polo shirts and our smelly sneakers, unmindful of how our imprudence might be dooming any one of us to lifelong incarceration in an iron lung and the realization of the body's most dreadful fears.

In Phillip Roth's long, descriptive, and dispassionate sentences, the story of a 1944 polio epidemic in Newark, New Jersey, relentlessly unfolds. Its dangers haunted every summer season between Memorial Day and Labor Day throughout the early decades of the twentieth century.

But summer camps were no safe haven, as Roth eventually reveals in his novel *Nemesis,* published in 2010. The dread of the disease was cruelly interlinked with all the pleasures of the holiday season. In 1947 Thomas Francis Jr., a National Foundation–funded professor at the University of Michigan, investigated an outbreak at a summer camp in Pennsylvania. Francis, highly respected as an epidemiologist, examined every detail with his characteristic care and thoroughness, but as to why a handful of children sharing the same food and drink, the same bathrooms, swimming pools, showers, and sleeping quarters went down with permanent paralysis while the great majority stayed fit and well, he was unable to discover.

The persisting mystery had little to do with a lack of funding for research. By 1940, the National Foundation had diversified its fundraising scheme. In addition to the Roosevelt Birthday Balls, local communities were laying out "Miles O' Dimes," while in movie theaters, powerful short films on the dangers of polio struck terror into audiences as March of Dimes mothers passed around collection boxes. In one, The Crippler became personified as a dark cloud menacing the summer sky, changing gradually into a frightful figure stooping to its deadly purpose over playground and sports field. As well as fear and terror, appeals to the sympathy and compassion of Americans were made by some of the best-loved movie stars. When Mickey Rooney bought an ice cream at the drugstore, he asked for a dime in the change and earnestly explained to young Judy Garland where he would send it, at which point Garland, brimful of anxious sympathy, added a dime of her own. Together they sealed their envelope to send it right away to President Franklin Roosevelt, the White House, Washington, D.C.

The movie theater collections raised hundreds of thousands of dollars every year. In 1938, funds collected by the National Foundation, also now know as the March of Dimes, reached $1.8 million. Then, in December 1941, Japan attacked Pearl Harbor and America mobilized for war.

What would happen to the March of Dimes? In terms of simple numbers, polio was not the highest priority as the nation contemplated the uncertainty, sacrifice, and hardship of what lay ahead. In 1941 there were 9,000 cases of paralysis, a modest number compared with what would happen in later polio epidemics. In New York City, Basil O'Connor and his team puzzled over what to do. Astutely, O'Connor decided to ask the president for his opinion. Roosevelt responded as O'Connor had hoped, by proclaiming the fight against infantile paralysis as one of the front lines of national defense. "Nothing," the president declared, "is closer to my heart than the health of our boys and girls and young men and women."

By 1945 annual contributions to the March of Dimes reached $19 million. Almost half came from local movie theaters. A powerful National Foundation cartoon by the African American artist Charles Henry Alston depicts a little black girl with her hair in braids, her arms spread wide in piteous appeal. A monstrous, shrouded figure with a skull-like face, *INFANTILE PARALYSIS,* stoops behind her, gripping both her legs. "Help me win *my* victory," she pleads, as victory in World War II draws closer. (Alston's bust of Martin Luther King Jr. would later be the first depiction of an African American in the White House.)

Despite the success of the foundation's fund-raising, the incidence of paralytic polio climbed relentlessly, though this was lost amid the hectic news of war. From 9,000 in 1941, annual cases reached almost 20,000 in 1944. No remedy was on the horizon. No breakthroughs had been made.

Of course, much of the foundation's funds were spent on helping those with polio. Everyone qualified for assistance, irrespective of financial means. Support was provided on a long-term basis but also in emergency responses to sudden epidemics. In June 1944, as temperatures climbed into the 80s, such an epidemic struck the town of Hickory, North Carolina, terrifying the people of the Catawba Valley. Despite the shock and fear the outbreak caused, the spirit of the people prevailed, helped by the emergency response of the National Foundation's local chapter. Foundation funds were made available to equip and staff a polio hospital. But even as the swimming pools and movie theaters closed, the playgrounds emptied, and the children were kept at home, the foundation was told that no such hospital existed within the quarantined area.

A call for volunteers went out, and a pine-board hospital was constructed by several hundred people, including county convicts, working around the clock on the grounds of a deserted summer camp. Three days later the hospital was ready, with everything provided by the local community. It was not a makeshift field hospital; it would become a fully functioning facility, with operating rooms, iron lungs, hydrotherapy tanks, a pathology lab, and specialized nursing and medical personnel. Several hundred patients were treated in the summer of 1944, many from outside the county. With the help of the National Foundation's public relations department, the event was destined to become "the miracle of Hickory." Still more funds were raised because of it, and they were sorely needed.

Franklin Delano Roosevelt died of a cerebral hemorrhage in April 1945 at the "Little White House" in Warm Springs, Georgia, 83 days into his fourth term. America mourned its longest-serving president. With the founder and greatest attractor of donations gone, Basil O'Connor fought hard to sustain continuity, stressing the foundation's vital work in the care of patients and the funding of research. Infantile paralysis was still gaining ground. In 1946, incidence topped 25,000, with no defense or remedy in the offing. What was needed was a breakthrough in research; a great leap forward.

Science does not proceed by collecting all the facts and then fitting them together to produce a scientific picture of life and the universe. Instead it begins with ideas. Based on their studies, scientists make up stories about what might be true and then proceed to test them out. As members of a world community, scientists aim to test and disprove these stories, searching for their shortcomings as consistent explanatory accounts of reality. In this sense, scientific knowledge is composed of all those theories that have yet to be discarded. Cold reason and the meticulous collection of data are important, but so too are unlikely chances, happy accidents, and luck. A good word for children interested in a career in science was invented by the English writer Horace Walpole in 1754. In a Persian fairy tale called *The Three Princes of Serendip* (modern Sri Lanka), the eponymous royals are forever making discoveries as they travel about, "by accidents and sagacity, of things which they were not in quest of . . ." From this Walpole derived *serendipity*—a very important means of scientific discovery. Louis Pasteur's cholera vaccine was discovered by serendipity; so too was that miraculous medicine, penicillin, which grew by chance on a culture plate left out over the holidays.

Nevertheless, given the misleading picture of polio as a neurotropic virus and the tragic consequences of two failed vaccines, a meticulous fact-finding era was deemed essential. This was inaugurated in 1938, when Basil O'Connor appointed Thomas Rivers, director of the Rockefeller Institute Hospital, to head the new National Foundation's committee on scientific research. Thomas Rivers was born on a Georgia farm in 1888 and went to Medical School at Johns Hopkins University in Baltimore, where he was diagnosed with a fatal muscle-wasting condition. He returned home to die but then changed his mind, setting off instead for Panama, where he worked as a surgeon. Here, his degenerative disease mysteriously stabilized, and he returned to Hopkins to complete his studies, going on to become a leader in the new field of virology.

Rivers's first task at the National Foundation was to identify research priorities. His list of topics was heavy on fundamental questions about how the virus gets in, how it is transmitted, and other basic aspects of viral biology that he judged should be tackled before a vaccine could be contemplated. To pursue these questions, scientists at Yale University and the University of Michigan, among others, received funds to investigate the transmission of polio (how it spread in the community), while Johns Hopkins scientists were funded to study how the virus spread and caused damage in the body.

The systematic implementation of this assault on the mysteries of polio owed much to the management skills of Harry Weaver, who was appointed as National Foundation research director in 1946. Weaver set new standards for grant administration and performance monitoring, which were later adopted by many institutions funding medical research. Among the primary questions framed by Weaver was: how many types of poliovirus exist? Two slightly different types were already known, and it was important to discover the full extent of variation—a vaccine would need to include all types. Weaver realized that hundreds of natural clinical isolates would need to be characterized in this way to find out for sure. In 1949, a young investigator at Johns Hopkins, David Bodian, together with his team defined three types of poliovirus, explaining for the first time why polio could sometimes strike the same individual more than once (one type did not confer immunity to the others). However, second attacks were thankfully rare.

Nineteen forty-nine was a good year for polio research. The most important breakthrough since Landsteiner's demonstration of a viral origin for the disease was published that year. The work began in 1948 and the magic ingredient—serendipity—played a part. In 1948, John Enders and his colleagues Frederick Robbins and Thomas Weller at Boston's Children's Hospital were trying to grow chickenpox virus in a mixture of human embryonic skin and muscle tissue in tissue-culture plates. "We had no immediate intention of carrying out experiments with poliomyelitis viruses," the three later wrote, but as they proceeded with their chickenpox studies, it suddenly struck them that "close at hand in the storage cabinet was the Lansing strain of poliomyelitis virus . . . everything had been prepared, almost without conscious effort on our part, for a new attempt to cultivate the agent [polio] in extra-neural tissues."

And they succeeded. The polio virus grew readily in the embryonic skin and muscle cells, and when they injected the cultured virus into monkeys, the animals developed the symptoms of polio. They later did the same thing for the two other polio types. Until this discovery, polio could not be produced in quantity in large-scale cultures because neural tissue couldn't be grown in tissue culture. But now the way was open for a vaccine. In 1954 John Enders, Frederick Robbins, and Thomas Weller received the Nobel Prize for their pioneering work.

Some of the mounting evidence that polio was not primarily a virus of the nervous system (as Simon Flexner had claimed) had come from the work of Albert Sabin. In a seminal study in 1941, he performed autopsies on polio victims and meticulously sampled tissue, showing that poliovirus was rare in nasal passages and olfactory nerves but plentiful in the digestive tract. It was clear from this study that polio arrived via the mouth and was primarily a gastrointestinal virus. If this was right then a vaccine should be entirely feasible.

At Johns Hopkins, David Bodian and Howard Howe took a different tack. They severed the nasal nerves of chimpanzees and then gave poliovirus via the mouth. The animals were readily infected, confirming the digestive tract as the route of infection and site of viral growth. Dorothy Horstmann, working at Yale, then found that when she fed chimpanzees the poliovirus, it appeared in the blood within days. David Bodian's results agreed with her findings. The new picture was highly encouraging for the development of a vaccine to induce protective antibody and defeat the scourge of infantile paralysis.

The field of polio research had made great strides. Toward the end of the forties, almost everything was in place for vaccine discovery. In fact investigators now had considerably more information than earlier vaccinologists had possessed when vaccines for tuberculosis, cholera, plague, typhoid, rabies, tetanus, and diphtheria were discovered. But given the false trails and the tragic consequences of earlier attempts, it was important to have reached this level of understanding. One crucial uncertainty remained: was *three* the full extent of polio serotypes, or were there others yet to be discovered?

The kind of work required to settle this question was dull and routine: repeating the same tedious assays on large numbers of patient samples. Weaver chose investigators he thought would do a thorough job to receive foundation funding. Among them was Jonas Salk, the young investigator newly established at Pittsburgh University Medical School. Salk already had grants from the U.S. Army for work on flu, but funding for the additional project was extremely welcome (as these things usually are). The money (just over $40,000) arrived in 1948, and Salk set his small team to work on the problem. As the funding increased over the next five years, he hired more people, and his laboratory tested clinical samples and autopsy material from all over the United States. The strains, each from an individual patient, varied in virulence. One isolate, called the Mahoney strain—originating from combined samples from three family members in Ohio—seemed most virulent of all.

The typing program, which ended in 1951, cost more than a million dollars. It turned out there were, after all, only three serotypes of poliovirus. As well as the financial cost, there was a high ethical price to pay. More than 17,000 monkeys were used in the assessment, which tested for signs of paralysis in monkeys made immune to one virus type and then infected with the others. The American Anti-Vivisection Society campaigned for legislation; religious groups in India, where many animals were sourced, raised concerns about appalling transportation conditions. By 1949, a scientist in David Bodian's department, Dr. Isabel Morgan, had shown that poliovirus could be inactivated with formalin (like tetanus and diphtheria toxins) and used as a vaccine in monkeys, protecting them against subsequent injections of virulent polio. And now the virus could be readily produced at large scale in non-neural monkey cells, which grew well in the culture flask. The scene was set. The final stages of the race to find a vaccine could begin.

But one intrepid investigator had already jumped the gun. Hilary Koprowski was not a part of the game plan orchestrated by the March of Dimes. Koprowksi was a graduate of Warsaw University Medical School, a musician and composer, a refugee from Nazi persecution, and a larger-than-life character oozing old-world erudition. Lederle, a pharmaceutical company in New York State (part of American Cyanamid), hired him in 1945. By 1947 he was injecting the Lansing strain (type 2) polio into the brains of mice, letting it grow, removing and mincing the brains, and injecting the liquidized material into cotton rats. By repeatedly transferring, or passaging, the virus through several groups of animals (more than 30 times), Koprowski succeeded in producing a weakened version, which he fed to nine chimpanzees. He then tried to infect them with virulent Lansing polio. All nine were protected. The next step was to test the vaccine's safety in humans. He and his assistant, Thomas Norton, drank the vaccine (liquidized cotton rat spinal cord containing attenuated polio). In a historical account of the work in the *British Medical Journal* in 1960, Koprowski recalled that it tasted like cod-liver oil. Both men remained well with no ill effects.

Koprowski then moved toward testing his vaccine in children. The laboratory director for Letchworth Village in New York's Hudson Valley, a nearby state institution for "the feeble-minded and epileptic," was a friend of Koprowski's and was anxious to test the candidate vaccine. His name was George Jervis, and he was concerned about the dangers of polio for children living in close quarters who were not able to manage rudimentary hygiene. For Jervis, the vaccine seemed a godsend. On February 27, 1950, Koprowski, Jervis, and Norton gave the vaccine to a six-year-old boy with no preexisting antibody to polio. The boy suffered no ill effects, and when his blood was tested fifteen days later, he had developed antibodies that neutralized the virus. The three men waited 44 days before feeding a second subject. After another favorable outcome, eighteen more children were given the vaccine, often in chocolate milk. No recipient suffered ill effects, and all developed neutralizing antibodies. "This," Koprowski wrote ten years later in the *British*

Medical Journal, "represented the first successful trial of immunization of man against poliomyelitis." He went on to say that the three of them kept the trial secret. Neither state officials nor senior managers at Lederle knew anything about the test. It remained secret until March 1951, when the National Foundation called a round table conference in Hershey, Pennsylvania.

All the important players were at this gathering: Thomas Rivers, David Bodian, Albert Sabin, the newcomer Jonas Salk, and John Paul, distinguished professor of preventive medicine at Yale. Paul, chairing a session, asked the pharmaceutical outsider, Koprowski, if he had any data to present. Koprowski, a big, deep-voiced, impressive man, then made his revelation. "The data I want to acquaint you with," he said, "represent a summary of clinical trials based on oral feeding of children with TN [Thomas Norton] strain of polio [living virus]." He went on to describe the results of his work at Letchworth. This sudden revelation was an enormous shock to the assembled experts who, until that moment, believed they knew everything that was going on in the small, elite world of polio vaccine research. It is a measure of Koprowski's gravitas and understanding of the paradigm that he held his own, arguing that his trial was in line with accepted practice. At that time, the use of institutionalized children in medical trials was the norm as long as three provisions were in place: that children were particularly suitable as subjects, that there was no discernible hazard to their health, and that they might directly benefit from the test. Permission from parent or guardian was also deemed essential. But Koprowski's main point, and his most telling in historical perspective, was that *someone* had to take this momentous step. Without such leaps into the unknown, Pasteur would not have given us his cholera vaccine and Jenner would not have discovered his vaccine against smallpox, which set in train the whole sustained endeavor to immunize against disease.

Koprowski went on to produce an attenuated strain 1 polio vaccine by a similar route, except that he took advantage of Enders's breakthrough and used tissue culture to produce his clinical material. But trouble lay ahead. A shake-up of senior management at American Cyanamid around 1954 led to a change in corporate goals and priorities and a new aversion to risk-taking. Koprowski, to say the least, did not fit well into the new ethos. But while contemplating his escape, Koprowski took advantage of an invitation from Queen's University, Belfast, to collaborate in a field trial of his vaccine. Sadly in 1958, during the early stages, virus recovered in stool samples from vaccinated children appeared more vigorous than the vaccine they'd been given, raising fears of reversion to virulence. The trial was immediately halted, and Koprowski left Lederle to become director of the Wistar Institute in Philadelphia.

The year following the revelations at Hershey was the worst in U.S. history for polio, but the figures (37 cases of paralysis for every 100,000 people) fail to convey the human tragedy. In September, a family living near Milwaukee was devastated by the disease. Four of their eight children were struck down by bulbar polio. In this, the most serious form of polio, the virus invades the cranial nerves that control breathing, swallowing, and speech. The tragedy is recorded in the *Milwaukee Journal* of April 13, 1955. The eldest, Paul, was affected first; an athlete of sixteen, he woke-up with headache, pain, and weakness in one shoulder. By evening, he could not cough or swallow. In hospital he was placed on a respirator at 6:30 p.m.; despite all the ministrations of intensive care, he died at 6:50. The next morning his four-year-old sister, Lorraine, woke with a headache and stiff neck and was rushed to hospital. Unlike her big brother, she ate well at suppertime, despite her sore throat, and fell soundly asleep, only to die without waking a few hours later. The day after this, her eight-year-old sister, Mary Ann, complaining of a sore throat and stiff neck, was rushed to hospital. When she began to vomit and had difficulty swallowing, doctors gave oxygen, penicillin, and plasma, and placed her in an iron lung. She continued to answer their questions until 6:15 p.m., when she died.

By now the Milwaukee family had lost three of their children, and they were praying hard for the remaining five. But two days later thirteen-year-old Barbara went down with a fever. Her headache was severe, she felt dizzy and nauseous, and in hospital she was fearfully aware of what her symptoms meant. Barbara went through the same intensive treatments as her sisters, but she died at 8 p.m.

The story of this family gives an inkling of the fear of polio, against which miracle drugs and the most recent advances in intensive care were powerless. Faced with this modern plague, the only hope, it seemed, lay in a vaccine. Hope, fear, and human sympathy fed the endless flow of dollars for the National Foundation's cause. In 1952, with the search for virus types complete, the foundation disbursed large amounts to investigators hoping to find a vaccine. Among them was Jonas Salk.

The big players in the world of polio research in America were relatively few. If an effective vaccine were to be discovered, it seemed likely that a handful of distinguished investigators, including John Enders, John Paul, David Bodian, Howard Howe, Thomas Francis, and Albert Sabin, would have a hand in it. But Jonas Salk had only recently arrived on the scene. Salk had presented his studies of viral typing at the Second International Poliomyelitis Congress in Copenhagen, Denmark, where the highlight was the news of Enders's breakthrough on culturing the virus. The journey home, on the great Cunard liner *Queen Mary,* had been organized by Harry Weaver. On board, Salk was introduced to Basil

O'Connor and his daughter Betty Ann, already a wife and mother of three. Salk's passion for his work, his sense of urgency, and his warmth and sympathy toward Betty Ann, a recent victim of polio, made a telling impression on the man presiding at the National Foundation.

Like Basil O'Connor, Salk came from a poor immigrant background and was the first of his family to go to university. Born on October 28, 1914, he was the son of Jewish immigrants from Russia who made their living in the garment trade. Jonas was the oldest of three boys, and his childhood was spent in East Harlem, the Bronx, and later Queens, as his family struggled up the social scale. At the age of twelve, he fulfilled his parents' sky-high expectations and entered Townsend Harris, a public high school for gifted children. Like many of the best from this school, he won a place at City College, which was the dream for any bright, ambitious immigrant growing up in New York City. Salk was about to turn sixteen. Tuition was free and the work ethic was intense. Beginning in the law program, where his grades were modest, Salk heeded his mother's advice and transferred to the premedical program, where his grades improved in everything but gym.

It was the Great Depression. Political consciousness, in a school where 80 percent of the student body came from the Jewish working class, was high; revolutionary politics, as traditionally practiced by students, was earnest and intense. And while it seems that the young Jonas Salk took no part in these activities, they began to shape the beliefs that would later make the man.

Salk's grades were just about good enough to get him into New York University College of Medicine. Unlike many of the top schools, such as Cornell, Columbia, and Yale, NYU did not employ a quota system to limit the proportion of Jewish students. Here Salk quickly came into his own, excelling at physiology, pathology, chemistry, bacteriology, and pharmacology. His energy and precocious ability were soon noticed by the dean, and in the latter part of his training, he was taught by Thomas Francis, a professor in the bacteriology department who was working on an influenza vaccine. In June 1939, Salk graduated and promptly married his sweetheart, Donna Lindsay. They had met at the renowned marine life laboratories at Woods Hole, Massachusetts, where Salk worked on his summer vacations. Donna was distinctly above him in the social scale, but eventually both families got used to the idea. Salk's new status helped—he won an internship, against the odds, at the prestigious Mount Sinai Hospital east of Central Park. At this time the hospital was leading the way in finding placements for doctors fleeing Nazi Germany. With the help of the National Committee for the Resettlement of Foreign Physicians, Mount Sinai was a home to a large number of émigrés.

It was during his time at Mount Sinai that Salk, strongly influenced by Donna, a fervent supporter of social justice, became active as a campaigner on political issues. He aligned himself with communist groups that supported the fight against fascism and campaigned against inequalities among the people of New York City; as president of the house staff of interns, he voiced vigorous support for the Allies in Europe. Such sentiments were fine at this time—the Soviet Union was an ally against Hitler, and left-wing radicalism was all the rage. But Salk's principled stand would not be forgotten when things changed in America. Toward the end of 1941, with his training coming to an end, Salk wrote to Thomas Francis, now chairman of epidemiology at the University of Michigan's School of Public Health at Ann Arbor.

This Midwest school was not Salk's first choice. He would have preferred to stay at Mount Sinai or join the Rockefeller Institute. But these prestigious posts were not on offer. Yet there were great advantages in having Francis as a mentor. Apart from his prowess in the field of influenza and vaccine research, Francis, as newly appointed head of the U.S. Army's Commission on Influenza, was able to intervene when the New York draft board identified Salk for service in the army. As "an essential investigator" in a field "of great importance to National Defense," the local draft board granted Salk a deferment.

The history of vaccination had already established live vaccines as the best way to stimulate a strong immune response. Jenner's vaccine was the pathfinder, and Pasteur's vaccines were also living microbes rendered harmless but still a big kick to the immune system. Sixty years on, with the need for a polio vaccine ever more urgent, many of the heavyweights believed a live attenuated vaccine was the only way forward. John Enders held this view, and so did Albert Sabin. But Francis was an expert on inactivated vaccines, and he was on the verge of developing a killed flu vaccine for the army. Francis's meticulous nature, together with his natural circumspection, was ideal for researching how to safely and consistently kill a dangerous virus. Cautious incremental advances were his style—the polar opposite of Hilary Koprowski's audacious pioneering. This was the culture in which Salk received his training in full-time research, and the style and substance of the work would shape his own discoveries. Two years after Salk joined the lab, Francis's formaldehyde-killed flu vaccine was undergoing trials, and the results looked good.

By the end of the war, as a research associate without security of tenure or prospect of advancement, Salk was on a modest annual salary of less than $5,000 and had a wife and child to support. During the next two years his salary went up a bit, and he traveled to occupied Germany to measure the success of the flu vaccine. But the time had come to break free and find his independence. In 1947 he made his escape.

Pittsburgh was not the obvious choice for a young investigator eager to establish a reputation. The city that produced half of all the steel in the United States had long been called the furnace of America, or "hell with the lid off" as the biographer James Parton put it. The hard history of immigrant workers in the steel industry, blighted by industrial strife and troubled by discrimination, had recently been chronicled by Thomas Bell in his documentary novel *Out of This Furnace*. As Salk arrived, strikes in coal and power industries were paralyzing the city; industrial pollution frequently darkened the skyline. And the run-down medical school was not renowned for scientific achievement. But a renaissance was about to begin, funded by the industrial magnates who'd grown rich on the city's industrial toil. The hiring of Jonas Salk was a small part of this vision for a new Pittsburgh.

Salk's start-up resources in the previously disused basement of the Municipal Hospital were modest, but his hopes were pinned on the future. Like many young investigators starting out on their own, he slowly began to build his scientific fiefdom. What Pittsburgh Medical School held for Jonas Salk was potential. With him he brought his army grant for work on influenza and a strong desire to work on polio. Infantile paralysis was easily the best-funded health challenge in America. The National Foundation saw to that. Harry Weaver organized the funds, and in 1948 Salk received the first check for his work on polio subtypes. As the years passed and the work progressed, the grants got bigger, until Salk's lab accounted for the great majority of external funding in Pittsburgh Medical School. Among the most expensive items were the monkeys—regrettably the only way to test for virulence. The work was routine yet essential. An effective vaccine would need to cover all types, but it assigned Salk a lowly position in the hierarchy of polio researchers. This was reflected in how he was treated at meetings, most significantly by the deeply patronizing Albert Sabin. According to the historian David Oshinsky, Sabin saw Salk as an errand boy for the National Foundation and a mediocre scientist; what would happen later did nothing to change his opinion. The typing work concluded in 1951 with no new subtypes to be added to the three already know: Brunhilde (type 1), Lansing (type 2), and Leon (type 3). The infamous Mahoney strain was a type 1 virus. (A strain is a particular clinical isolate which may be propagated in the laboratory. Each strain falls into a particular viral type. While there are only three viral types, each type contains multiple strains.)

No longer the struggling postdoctoral researcher, Salk was now an established investigator. Given the flawless conformity of his new life in Pittsburgh, a Federal Bureau of Investigation file on his former left-wing activities was left to gather dust. Salk now lived in the suburbs with his wife and three small boys—a full professor running a top-class virus research lab at the age of 36 and commanding a salary of $12,000 a year. From the security of this well-funded, fully established position, and with the extent of viral diversity now confirmed in his laboratory, Jonas Salk was ready to move forward with the work he desperately wanted to do: the discovery of an effective vaccine against polio.

14

A GREAT STEP FORWARD

"But the first thing you notice about Christina is her voice. It is firmly lucid, a practical countrywoman's voice, and it immediately wipes out her myth, the romantic, half-pitiable aura of a spirit straight-jacketed into immobility."

Brian O'Doherty, *Show V,* No. 4, May 1965

By 1950 Jonas Salk had already tested both live attenuated and killed polio vaccines in monkeys, the latter using the formaldehyde inactivation already developed for flu; he'd found them effective, stimulating good levels of protective antibody. He was now ready to begin testing in humans, and he confided this to Harry Weaver at the National Foundation. Hilary Koprowski's pioneering work was still a well-kept secret, and no one would learn of it until the National Foundation committee meeting of March 1951. Weaver was cautious but indicated, implicitly, that a grant application to fund this monumental step would receive the most careful consideration. Salk's team had already succeeded in producing virus in tissue culture using monkey kidney cells, and he now proceeded to select the best strains for a vaccine. Salk's long familiarity with formaldehyde treatment made him confident that even the most dangerous viruses could safely be included: however virulent, they could, with certainty, be killed. For type 1, responsible for 80 percent of human polio, he chose the notorious Mahoney strain. Another two strains covered the remaining types. Each was inactivated separately using the conditions worked out at the University of Michigan and fine-tuned at the University of Pittsburgh School of Medicine. With precisely the right amount of treatment, the virus could be killed but its structure preserved. An adjuvant in the form of mineral oil, known to enhance animal immune responses, was also included. To establish vaccine safety, monkeys received injections directly into the brain, were observed for a month, and then autopsied for signs of damage or disease.

In December 1951, an expert committee convened by Harry Weaver considered Salk's proposal for a human trial. The panel included David Bodian, John Enders, Thomas Francis, John Paul, Tom Rivers, Albert Sabin, and Salk himself. Salk did a good job on his presentation, and the team then argued passionately about the merits of the plan. The biggest hitters, Enders and Sabin, believed that the only possible serious contender for large-scale vaccination was a live attenuated virus. Francis and Salk were outgunned.

The sense of urgency at the National Foundation could hardly have been greater. Their fund-raising promise, that a vaccine was imminent, had already been made to the American people. In 1949 they'd announced that the conquest of polio was in sight. The pressure to deliver was intense and the decision makers knew it. The foundation's expert committee could not agree, so what could be done? Ignore the expert panel?

The decision-makers at the National Foundation, O'Connor, Rivers, and Weaver, quickly decided they *had* to move forward, and they would do so in secret. Salk had already done some vital groundwork: two institutions, the D. T. Watson Home for Crippled Children and the Polk School for the Retarded and Feeble Minded, both conveniently close to Pittsburgh, were willing to volunteer the children in their care. The State of Pennsylvania approved the plan—the disabled children at Polk, living in close quarters and unable to manage their personal hygiene, were at elevated risk of polio and stood to benefit from a successful vaccine.

Testing began in June 1952. The Watson Home was a leader in polio rehabilitation, and one of the children, Bill Kirkpatrick, courageously volunteered to go first. Bill, a keen football player, sixteen years of age, had been struck down with polio over Labor Day weekend the previous year after training hard on the running track. Testing in a child already paralyzed and with existing antibody was the safest way to start. Everything went well: Bill was none the worse for the injection and his antibody levels rose. Forty-two other young victims of polio were then injected, most of them by Salk himself, with similar results. It was time to move on to Polk, where many children had no preexisting immunity to polio.

Polk was a very different institution, with overcrowded wards and a chronic shortage of staff. On these handicapped children Salk tested his single and his mixed triple-strain vaccine. He also tested vaccine with and without the mineral oil adjuvant. He was, of course, deeply affected by these trials. He knew all too well that the history of polio vaccination was not a happy one. But the vaccine was safe—no children suffered ill effects, and they produced good levels of antibody. Salk was elated. Nothing that subsequently happened in his scientific career would ever compare with that unique thrill of discovery in the winter of 1952.

The pilot trial could not have been more timely. In American history, 1952 was the worst year for polio. By the year's end more than 20,000 had been paralyzed and 3,000 had died. The top man in charge of combating this terrifying plague suffered a heart attack in June, but after three months recuperation, the indefatigable Basil O'Connor came back fighting. At a meeting of the National Foundation Immunization Committee in Hershey, Pennsylvania, on January 23, 1953, Salk presented the results of his trial on 161 children. This total now included his three young sons.

The stormy response that followed did not quite equal that provoked by Hilary Koprowski two years earlier, but it wasn't far off. Fears about safety and Salk's choice of the fearsome Mahoney strain stoked up tension in the room; also, the use of mineral oil adjuvant was, to say the least, controversial. Joseph Smadel of the Walter Reed Army Hospital, pointing to the serious consequences of delay, spoke up in favor of a major trial, but others were vehemently opposed. Those against included the heavyweights—John Enders, Albert Sabin, and Howard Howe believed that much more research was needed and that only small pilot studies could possibly be justified at this early stage. A big trial of a less-than-perfect vaccine could have disastrous consequences. Weaver said nothing. The National Foundation had already disregarded the committee and could, if they so decided, take an even bolder step.

The events that followed illustrate the volatility that sometimes comes with breakthroughs of this kind. Whether or not it really was a breakthrough would be decided not just by the experts but also by the American public. Two small steps ensured this would happen. In contrast to Koprowski's extraordinary secrecy, Salk leaked the news of his Watson and Polk trial to a trusted reporter. This journalist remained silent until Harry Weaver, for reasons of his own, strongly hinted to the National Foundation's board of trustees that a very promising new vaccine was in the pipeline. Once this was in the public domain, the journalist felt free to break the story, and on February 9, 1953, *Time* magazine carried the news. Albert Sabin and the other experts were appalled; Sabin let Salk know in the most withering terms. But Basil O'Connor, enlisting the support of Tom Rivers, quickly arranged a formal briefing of the press. Some of the resulting national headlines had little to do with scientific caution.

What followed was a little strange. Salk, complaining of pressure to proceed at once and indicating that he needed more time to fine-tune his vaccine, agreed to a personal appeal on national radio. It took place on the evening of March 26. His stated aim was to calm things down, but not surprisingly the broadcast had the opposite effect. While saying that he needed one more year, he also effectively announced to the nation that he had a vaccine that worked. Two days later Salk's report on preliminary human studies with his vaccine appeared in the *Journal of the American Medical Association*. A few weeks after that, O'Connor convened a brand-new committee to advise on designing a large-scale trial. It included no one from the old and intransigent Immunization Committee.

A highly respected vaccine expert at the National Institutes of Health, Joseph A. Bell, was appointed to direct the design of a large-scale trial. O'Connor and the National Foundation had favored a volunteer trial with injected and noninjected schoolchildren who could then be monitored over the following year's polio season. The outcome would be measured by the incidence of paralytic polio in vaccinated and unvaccinated groups. Since polio only paralyzes a very small proportion of those infected (less than one in a hundred) many thousand children would be needed to ensure a meaningful result. But Bell wanted to be more rigorous—he proposed the inclusion of injected controls: groups of children receiving injections containing no active vaccine. Moreover, he stipulated that the trial must be "double blind" to avoid any bias: neither the child nor the vaccinators must know whether vaccine or placebo (mock vaccine) had been given until the results were in. A clinical field trial on such a grand scale was completely unprecedented. As the plans crystallized and the public, so long in dread of polio, dared to hope, Albert Sabin publically declared his lack of faith in Salk and his opposition to the trial, driving home his message that the only way to prevent polio was with live attenuated vaccines.

Bell also made it clear that the mineral oil adjuvant should be dropped. It had in the past been implicated in adverse reactions—swelling and troublesome sores—and there was concern about the risk of tumors. Salk conceded on this point even though, without the extra kick, three injections, not one, would probably be needed. But he refused to budge on the mock injection issue, declaring any mock-injected child subsequently paralyzed would weigh too heavily upon his conscience. O'Connor sided with Salk—he envisioned an uncomplicated trial the public could welcome and embrace. The complex logistics of a double blind, placebo-controlled study troubled him deeply. With the need to end widespread fear and individual tragedy so urgent, why complicate matters? Double blind trials, in which those giving and those receiving the vaccine do not know if it is active or an inactive placebo, are the best way to eliminate subjective bias on the part of both experimental subjects and experimenters. This provides the highest level of rigor in obtaining a true picture of vaccine effectiveness. The facts about what was actually given in each case are not revealed until all the results have been processed.

But Harry Weaver did not agree with the opinions of his chief. The gifted, dedicated administrator had appointed Bell precisely because of his scientific gravitas, and he stood behind the rigorous approach. So too did Tom Rivers: nothing could be worse than failing to achieve a convincing result. An emotional debate quickly got underway, and when O'Connor seemed to sideline Harry Weaver in his dealings on trial design, Weaver resigned.

It was August 31, 1953. The Cold War was deepening. On August 14 the Soviet premier, Georgi Malenkov, nonchalantly announced momentous news: the Soviets too had succeeded in producing the hydrogen bomb. Soon afterward, the two greatest fears reported by Americans were a nuclear holocaust and the scourge of polio. But just as public demand was growing, the campaign against infantile paralysis seemed to falter. Sabin condemned the Salk vaccine as unready for mass testing, fervently arguing the need for more research and the folly of haste. And in September Bell's tentative plans for a double blind vaccine field trial including placebo were effectively rejected by the foundation, precipitating his resignation on October 1.

But apart from Sabin, the other key opinion leaders—Bodian, Paul, and Howe—by now felt differently: at this stage they believed it was just too late to pull back. Thomas Francis, careful, polite, but meticulous in his criticisms, was Sabin's most entrenched opponent, implying that opposition at this late stage could only be damaging to all. What Sabin didn't know was that Francis had already been asked privately to oversee the Salk vaccine trial, the largest field trial in history.

Only when complete autonomy in the conduct of the trial had been granted him did Thomas Francis agree to take it on. Famed for his rigor and scrupulous attention to detail, Francis reinstated the double-blind component, in which a large group of randomly selected children would be injected with placebo. Neither the children nor the investigators would know which children had received the real vaccine and which had received the mock vaccine. The standard had been set by other trials and nothing less would do, he insisted. O'Connor acquiesced at last, while Salk, who retained great respect for his mentor, finally agreed as well, surrendering what once had seemed a heartfelt principle.

Because the incidence of disease was so low, more than a half-million children would have to be injected in order to establish a statistically significant result. Several hundred counties were identified with sufficient school-age populations and higher historical rates of paralysis. Good links with National Foundation local chapters, to aid the volunteer part of the operation, also shaped the final choice. Including the observed controls (age-matched children receiving no injections) almost a million and a half children would take part. The scale of the exercise was reminiscent of national mobilization: over 600,000 youngsters, ages seven and eight, would receive the three separate injections over an eight week period. Clinical trials had come a long way since the tests of variolation on Newgate prisoners and the orphans of London's St. James more than 200 years before.

There was an air of national excitement. By this time, a majority of Americans had donated to the March of Dimes. The National Foundation had both the power and the enormous responsibility: in line with the dread of socialist medicine that characterized the times, no federal dollars or public money was involved. Melvin Glasser, the man charged with coordinating the whole endeavor, faced a task that required extraordinary planning and execution. These days we have computers to model the logistics of large operations, but not back then: tens of thousands of professionals and volunteers were trained for the job. Persuading parents to volunteer their children was no great problem: families were keen to be part of one of the most important developments in medical history, and the chance to be included was cast as a privilege.

To produce sufficient vaccine, foundation commercial contracts were arranged with Connaught Laboratories in Canada and Park Davis in Detroit. After various technical problems, six other pharmaceutical companies were added to avoid the possibility of a shortfall. Some, like Wyeth and Eli Lilly, were large; others, like Cutter Laboratories, were relatively small. However, Park Davis and Eli Lilly quickly produced almost all the vaccine required for the trial.

Under the lens of Francis's scrutiny, old fears and new concerns kept surfacing. Sabin re-iterated his objection to the inclusion of the virulent Mahoney strain. Others worried about validating the inactivation process: how could you be sure that every particle of virus had been killed? Still others harbored fears of hidden viruses acquired from monkey kidney cells used in production. On April 4, 1954, on the brink of the trial, the press ran stories of live poliovirus sneaking through and causing disease in the monkeys used in the test process. There was substance to the claim: lot testing in monkeys had detected pathogenic batches produced by the scaled-up process. David Bodian was called in to investigate, and suddenly the whole endeavor was in jeopardy. Bodian conducted a careful analysis of the situation and insisted that additional testing be added to prevent defective batches getting through.

Testing now consisted of triple safety checks in which companies had to produce at least eleven consecutive clean lots before any lots were passed. Even one failure (even one batch causing sickness in the test monkeys) meant that all eleven would be failed. On this strict basis the green light was on again in the nick of time. By now Salk had accumulated more than 5,000 vaccinated individuals, and the fact that there were no untoward effects was further reassurance for parents committing their children to the trial. On April 26 Randy Kerr, a six-year-old boy at Franklin Sherman Elementary School in McLean, Virginia, bravely and cheerfully received the first shot.

An operation on this scale was bound to have irregularities. Some vaccine got mixed up with placebo, some was stolen, some misdosed, and some mislaid. Elsewhere everything proceeded smoothly according to local convention: in Montgomery, Alabama, black children, summoned by their first name only, assembled for their injections outside the walls of whites-only public schools and were not allowed inside to use the washrooms. With over a million children taking part, several hundred deaths occurred as fortune and misfortune took their usual toll. The only difference from normal was the extreme scrutiny: each time a child died, Tom Francis took an urgent phone call. Accidents, cancer, respiratory infections, and polio, of course, all played their part. In the blinded placebo-controlled trial, paralysis was a necessary part of the process. Its incidence was all-important, and the crucial difference between vaccinated and control group would decide the fate of Jonas Salk's vaccine.

By the end of June, the vaccinations were completed according to plan. It was an extraordinary achievement. Schools closed for summer as the polio season arrived. As the weeks rolled on, an army of workers swarmed through the Vaccine Evaluation Center in Ann Arbor, Michigan, an old hospital building adapted for the purpose. So much was at stake, but nothing could be hurried—the American people would simply have to wait. Would the vaccine work? No one could be certain. Its efficacy in preventing disease had never been tested in humans before. Salk's vaccine was a killed vaccine, entirely unproven, and certain leading experts had no confidence in it. What's more, no adjuvant had been included—that magic extra kick to the immune system. Could three doses possibly compensate for this? But a fourth concern troubled Jonas Salk. At the last minute National Institutes of Health representatives had demanded the addition of a preservative, Merthiolate, to guard against bacteria and molds. Salk had evidence that this compound compromised the effectiveness of his vaccine.

It was not until March 1955 that the inscrutable Francis was ready to write his report. It would take, he said, about a month. The results of the 1954 field trial of the Salk polio vaccine would be announced in the auditorium of Rackham Hall, the graduate school of Michigan University at Ann Arbor on April 12. Salk and O'Connor first learned of the result over breakfast with Francis that very morning. The American public, represented by a press corps 150 strong, would not have to wait much longer. At precisely 9:10 a.m., each reporter received his or her own summary of the Francis report. When Thomas Francis had completed his public address, they would be free to broadcast the news.

But it was just too much to ask. Only ten minutes later the verdict was out. Or rather, the press corps "take" on the verdict was out, just in time for NBC's *Today Show*. All across America people wept and cheered. As the deeply serious, highly principled chief evaluator of the biggest clinical trial in history rose from his seat to begin his measured and dispassionate account in Ann Arbor, jubilation swept the nation. Car horns blared, church bells rang. The vaccine was safe, effective, and potent. Polio had been defeated!

Inside the Rackham Auditorium, the audience was silent as Thomas Francis calmly delivered his address. He spoke continuously for more than 90 minutes, during which the distinguished audience gradually digested his carefully considered elaboration of the clinical field trial results. Comparing vaccine- and placebo-injected groups, the most rigorous part of the field trial, the Salk vaccine was 80 to 90 percent effective in preventing paralytic polio. This was the average of its 60 to 70 percent effectiveness against type 1 virus and its 90 percent or more effectiveness against types 2 and 3. But comparing vaccine and observer (uninjected) groups, the Salk vaccine was 60 to 80 percent effective (60 percent against type 1 and 70 to 80 percent against types 2 and 3). From the outset, Francis had worried that the latter comparison was unfair. Volunteers for injection tended to come from advantaged families—a group at higher risk of polio. Observer controls included children from less advantaged communities—a group known to be less at risk from polio. The true effectiveness of the vaccine was likely to be obscured by this comparison. Despite all this, the Salk vaccine *was* safe and *had* proved effective, if not quite as miraculously effective as the jubilant media circus sang out to the nation. While the *New York Times* went with "Lasting Prevention of Polio Reported in Vaccine Trials," other newspapers were less restrained. "Salk Vaccine Whips Polio," proclaimed the *Light* in San Antonio; "Polio Routed!" was the *New York Post*'s headline.

Now the public clamored for the vaccine, and the government rapidly responded. The Salk vaccine was quickly licensed by the newly created Department of Health, Education, and Welfare, and thanks to Basil O'Connor's confidence before the outcome of the trial, by mid-April

1955 six commercial manufacturers already had sufficient vaccine to begin distribution.

Salk was heralded as a hero. The American public adored him, and the unassuming, shy demeanor of the man they heard and saw through the media only increased their adulation. But Salk's peers thought otherwise. In his speech following Francis's talk, Salk failed to give credit to his team, several of whom had made crucial inventive contributions to the project. Instead he spoke at length of how he had improved his vaccine still further, an emphasis in danger of detracting from the trial's achievements. And his peers well knew that the vaccine entailed no new discovery or innovation. Instead it was an extension of already established principles and methodologies. When the National Foundation and the University of Pittsburgh reviewed the possibility of patents, Salk himself pointed out that securing patents on inventive steps would be difficult—his vaccine was developed from the ideas and techniques of others.

Scientists who court the media always risk the disapprobation of their colleagues, and some viewed Salk as a clever manipulator of the press. But without him—without Salk's pragmatism, his single-minded drive, and his unflagging sense of urgency—there would have been no polio vaccine for another decade.

On his second appearance on the cutting edge TV program *See It Now,* hosted on April 12 by the distinguished and respected broadcaster Ed Murrow, Salk delivered his most famous and effective line to the American public. Murrow began by describing the polio vaccine as a "giant step forward," which had lifted a sense of fear from the homes of millions of Americans. He went on to ask Salk who owned the patent on the vaccine. Salk replied that "the people" owned the patent, adding, "There is no patent. Could you patent the sun?" His place in the hearts of the American people was assured forever.

As Salk was showered with civic honors and thanked personally by an emotional President Dwight D. Eisenhower in a ceremony at the White House; as the delivery of vaccine stuttered into action across the nation, aided by a federal government at first reluctant to interfere in the domain of private enterprise; as letters of heartfelt thanks from parents across the nation flooded into Salk's laboratory, a terrible tragedy was brewing in the West. Early on Sunday, April 24, a little girl in Idaho came down with symptoms resembling polio. She had received the Salk vaccine six days earlier. Officials assumed that, as with many tragic cases in the great field trial, she had received the vaccine too late, or it hadn't sufficiently protected her. She died three days later. By this time four more polio cases had been reported to state officials. It was not yet the season for polio, particularly in Idaho, a Rocky Mountain state due east of Oregon. Soon the number of cases in recently vaccinated children was approaching a dozen. Reports came from Chicago down to San Diego. All, it seemed, had received Salk vaccine produced by the same company—Cutter Laboratories, a family-run pharmaceutical house in Berkeley, California.

At an emergency teleconference linked together in the small hours of the night, a handful of top experts urgently conferred. Thomas Francis and Joseph Smadel were among them. Complaining of very limited evidence, the anxious specialists could not agree on halting the use of Cutter vaccine. Several hundred thousand doses had already been given, and several hundred more were imminent the very next day.

On April 27, with new cases mounting, U.S. Surgeon General Leonard Scheele acted to halt the use of Cutter vaccine with immediate effect, dispatching an investigation team to Berkeley, while the Public Health Service urgently organized surveillance of cases among recently vaccinated children. Was Cutter the only vaccine to be incriminated, or was the catastrophe more widespread? On May 8, Scheele halted *all* polio vaccination and went on national TV to justify his decision. Vaccine from other producers had not been incriminated, he explained, but caution was essential pending a full investigation.

The press made much of the turmoil. Basil O'Connor blamed both government and commercial producers, while Salk stood by the safety of his vaccine when produced in strict accordance with his protocol. And President Eisenhower seemed to speculate that government testers, faced with huge demand, might have taken shortcuts. The elite virologists who'd always preferred a live attenuated vaccine lamented the hijack of scientific endeavor by public opinion, leading to such haste and opening the door to disaster. Albert Sabin, silent since the field trial results, now went public, condemning all killed vaccine that contained the virulent Mahoney strain.

On May 13, with the polio season drawing ever closer, Surgeon General Scheele was able to clear certain lots from Park Davis and Lilly after rigorous safety testing. Parents who had clamored for the vaccine now felt deeply anxious and unsure. The distinguished panel of experts assembled by Scheele to testify at a special congressional hearing was predictably divided. Opponents of the Salk vaccine, led by Albert Sabin, were adamantly against proceeding—time was needed to develop a safe and effective live virus vaccine, as this tragedy so clearly confirmed. Supporters pointed to the folly of waiting for a perfect vaccine, as there would always be some further improvement to be made. They took a vote. Eight were in favor of proceeding with national vaccination; three (Sabin, Enders, and William Hammond of Pittsburgh University) were against; the four remaining experts (including Salk) abstained. The vote to replace the Mahoney strain as soon as this could be technically achieved was easily carried.

So the campaign went ahead. Relatively few children received their shots in the summer of 1955, and major epidemics afflicted Boston and Chicago. The dark cloud of "The Crippler" once more hung low upon the holiday season. More than 200 polio cases were traced to six Cutter vaccine lots. Seventy-nine of them were vaccinated children; the rest were family members or community contacts. Most suffered severe paralysis. Eleven victims died.

At his lowest point in early May 1955, Salk felt suicidal. The Cutter tragedy was taking its toll and the Nobel Prize winner in medicine of that year, John Enders, condemned his vaccine as quack medicine. Investigations eventually implicated settling-out in production cultures as the most likely cause of live virus getting through. In this event, clumps of virus are formed in the sediment that gathers on the floor of culture vessels left to sit too long; the chemical used to inactivate the virus then fails to penetrate to the center of these clumps so that live virus, protected by the killed virus surrounding it, persists. Moreover, it seems the filters used to remove such clumps were less consistent than those used by Jonas Salk. Lower-quality filters lacked the narrow range of pore sizes required to exclude all unwanted particles and may have allowed some clumps to pass through. Today, there is a complex bioengineering science surrounding the correct agitation and mixing of scaled-up biological cultures. These measures, which employ high-tech rotors, or impellers, prevent sedimentation and ensure an even and equal exposure of living particles (be they viruses, bacteria, yeasts, or cells) to the molecules in the culture medium. But this science, which began in the brewing industry, had not yet been developed.

The Salk killed vaccine against poliomyelitis proved safe thereafter, and Salk's achievement was celebrated around the world. Following the mass immunization campaign promoted by the March of Dimes, the annual number of polio cases plummeted from 35,000 in 1953 to 5,600 by 1957. By 1961 only 161 cases were recorded in the United States.

The new science of virology was also making progress. Albert Sabin had been working on a live virus vaccine ever since the revelations of Hilary Koprowski's pioneering human trial in 1951. As early as the winter of 1954, Sabin was testing his attenuated vaccine in humans. Thirty adult prisoners at Chillicothe Federal Penitentiary in Ohio, conveniently close to his laboratory at Cincinnati Children's Hospital, were selected to receive the vaccine. Almost the entire prison population had volunteered to take part. The modest incentives ($25 and a little shaved off of sentences) were probably unnecessary—the tedium and isolation of confinement usually prompt prison inmates to participate in this kind of program. The prisoners selected had no existing antibodies to polio. The trial went well: good antibody levels to all three types of virus were raised, and not one prisoner took sick.

Sabin could be reasonably confident his candidate vaccine was safe in humans. The three weakened viruses had produced no signs of disease when injected directly into the spinal cords of laboratory monkeys, and this was the expected outcome given the way in which they'd been produced. Sabin began with the laboratory polio strains long propagated by other investigators; Hilary Koprowski was among those who sent him strains. As his starting point, Sabin included virus originally derived from the notorious Mahoney strain he'd so long reviled. But this virus, and two others representing types 2 and 3, had by now undergone numerous changes. By passaging, or transferring, the virus every 24 hours, dozens of times, into cultures of fresh non-neural tissue cells from monkey kidney and testis, Sabin derived strains of virus that were non-neurotropic; in other words, they were unable to grow in nerve tissue. This and other spontaneous changes in the virus's genetic material meant that it was rendered harmless. The aim was to produce stable virus that could grow in cells of the human digestive tract but would not invade the nerves or revert to virulent behavior when recultured from fecal material. This last property was crucial, and some early vaccine candidates had just this problem. Because Sabin's attenuated viruses grew naturally, the immune system saw them as active invaders and mobilized wider immune defenses than it did against the killed Salk vaccine.

With his best candidate triple vaccine, Sabin had already immunized himself, and following the success at Chillicothe, he was ready to test more widely. But there was a problem. His sponsors, the National Foundation, already had a safe vaccine that was working well and rendering millions of children safe from polio. They had little incentive to test an untried competitor. The solution came from abroad. Eisenhower had pledged to share American know-how on defeating polio, and despite Cold War tensions a Russian delegation arrived in 1956 to look at Salk's vaccine and the live-virus alternatives. Sabin was quick to indicate his willingness to reciprocate and visit Russia. After clearance from the FBI, Sabin flew to Leningrad that June.

Russian interest in the potential of the Sabin vaccine was high, not least because of serious epidemics of polio the previous year, and Sabin established a strong relationship with Mikhail Chumakov, director of Moscow's Polio Research Institute. Following some difficulties in producing and using the Salk vaccine successfully, the Russian authorities embarked upon a vast immunization program with Sabin's product. The three live viral types were given separately by mouth, a month apart. In the totalitarian state, compliance was readily achieved. In 1959, 10 million Russian children were vaccinated in a regimented, meticulously documented trial without mock vaccination or observer controls.

As the data accumulated it became clear that Sabin's vaccine strains were both safe and effective, and without more ado the Russian authorities moved rapidly to institute a program of national vaccination for every Russian under twenty years old. The massive program to inoculate more than 70 million children was soon underway. Sabin was not so carried away by success as to forget the importance of international scrutiny and verification of the results. He pressed for independent observers, and John Paul was quick to involve the World Health Organization.

Dorothy Horstmann, distinguished for her observations on polio viral pathology, was duly appointed to observe the Russian campaign and produce a report. Though she complained about the difficult logistics of such a vast trial and the problem of collating such a mass of data, her interim conclusions were favorable. The vaccine appeared to be working. Russian success was confirmed in presentations by Soviet delegates at the Second International Conference on Live Polio Vaccines, which took place in Washington, D.C., in June 1960.

In August that year, the *Journal of the American Medical Association* remarked editorially that small pilot studies were too limited and mass vaccination studies too large to accurately determine live vaccine efficacy. The article pointed to the value of an intermediate study of 850 children, carried out without fanfare in Philadelphia with very good results. This, they hoped, would convince those still reluctant to receive the Salk inoculation to swallow instead the live attenuated vaccine. A large trial in the United States, the journal opined, was now called for.

When this article appeared, such a trial was already underway. Though the vast Russian program had put Sabin in the lead, there *was* another live vaccine in the race. Harold Cox at Lederle Laboratories had by now revived the legacy of polio's live vaccine pioneer, Hilary Koprowski, and the company was developing a triple vaccine. In contrast, Sabin's three viral types were still given separately. Relations between Cox and Sabin were strained to say the least.

Sabin's trials began in Cincinnati and surrounding Hamilton County, Ohio where 200,000 children received the vaccine on sugar lumps or in syrup. If Salk had subtly steered the fortunes of his vaccine, Sabin was more direct: he vigorously prosecuted the conduct of the trials. Meanwhile, Cox proceeded with his single-dose triple vaccine in Dade County, Florida. Four hundred thousand received this vaccine. What happened next would decide the winner. A half-dozen polio cases among Florida vaccinees caused concern, which tipped the balance in Sabin's favor. A causal relationship between Cox's vaccine and these cases was never established, but the uncertainty was enough. In August 1960, the surgeon general approved trial manufacture of Sabin's vaccine.

The Salk vaccine was working well. In 1961, the last year in which it was exclusively used in America, polio cases dropped below 1,000. But perceptions of its effectiveness had already been influenced by small resurgences of the disease in 1958 and 1959. With polio no longer the great evil to be feared alongside nuclear war, the incentive to get children vaccinated had been considerably reduced. The course of three Salk injections also made for lack of compliance, especially in disadvantaged groups where rates of vaccination were relatively low. Would the introduction of a new vaccine, easily swallowed, revive the vaccination rate? Sabin lobbied hard, and rifts developed at the National Foundation over what best to do. Some asked why Russian children could benefit from the new product when American children could not. Then, in 1961, the American Medical Association demanded clarification on which vaccine to use. The resulting investigation, unduly influenced by drug company enthusiasm for the new product, recommended that the Salk vaccine be replaced as soon as the Sabin vaccine was generally available.

All Salk's protests were in vain. In September 1961, the Department of Health, Education, and Welfare licensed the Sabin vaccine as the replacement for the old killed vaccine. Large-scale vaccination began in 1962, and polio incidence continued to fall. In 1979, the last indigenous case of polio was recorded in the United States.

In 1963 Jonas Salk founded the Salk Institute near San Diego, California. This advanced research institute is housed in a dramatic, purpose-built building around a grand courtyard open to the ocean, and its scientists are free to pursue their individual interests in fundamental biological research. Salk wryly observed that he could never have hoped to gain membership of such an elite institute had he not founded it himself.

Sabin was never taken to the hearts of the American people as Salk had been and would continue to be. The rivals were of similar backgrounds: both men were from working-class, Jewish immigrant families. Sabin arrived with his parents, age sixteen, from Bialystok in Poland in 1921, escaping persecution of the Jews. But while Salk appeared to be an unassuming, ordinary man who'd struggled to rise in order to dedicate himself to the defeat of disease, Sabin always seemed remote: an elite scientist immersed in the mysteries of his field. But Sabin's vaccine did possess important advantages, not least the simplicity and acceptability of being swallowed. But this is not the only benefit.

Poliovirus enters the body through the mouth, usually in food or water contaminated by feces, or through physical contact. All those infected shed virus in feces for several weeks, but in most cases (90 percent), the virus is limited by the immune system, reproducing in gastro-intestinal cells and circulating in the blood, where it causes no symptoms and no after-effects. In 5 percent of cases, so-called abortive polio occurs 3 to 21 days after infection: a slight fever and sore throat develop together with vomiting, headache, and abdominal pain. The illness only lasts two to three days. In about 1 percent of cases, the signs of abortive polio are present, but the headache, nausea, and vomiting are much worse. There may also be stiffness of the neck, trunk, and limb muscles. This is called nonparalytic polio. Paralytic polio occurs in about 0.1 percent of cases, when spinal cord nerve cells are affected. By damaging motor nerves in the spinal cord, poliovirus causes a floppy (or flaccid) paralysis of muscles in the limbs. Commonly, just one limb, a leg or an arm, is paralyzed, but paralytic polio may affect groups of muscles in more than one limb. When brain stem nerves are damaged in bulbar polio, breathing, swallowing, and bladder and bowel function become compromised. The more severe the disease, the more the patient is likely to die.

Paralysis may improve over six months, but many are left with long-term disabilities. Strenuous physical activity during exposure to the virus *does* appear to increase the chances of paralysis. Because it enters by the natural route and actively grows in cells, Sabin's attenuated polio vaccine is a more potent stimulus to the immune system than killed vaccine, and it produces longer-lasting protection. The antibodies raised by the live vaccine neutralize virus entering the digestive tract as well as in the blood. In contrast, Salk's vaccine raises antibodies largely confined to the blood, where they don't attack the virus until it reaches the circulation. Because the Sabin vaccine grows in the digestive tract, the attenuated virus is shed in feces and can immunize those in the community who happen to encounter it by chance. There is however, a significant disadvantage to the Sabin live-polio vaccine: in rare cases it reverts to virulence and causes paralytic polio in around 1 of every 2.4 million vaccine recipients.

Today, it is Sabin's live attenuated vaccine that is being used in Africa and Asia in the global endeavor to eradicate polio. And Sabin's vaccine also triumphed over Salk's in America. Though he lived until 1995, Salk never saw his vaccine in large-scale use again. But the story did not end there.

With polio now a part of history in the developed world, how are its victims perceived? There are around 400,000 Americans living with the effects of polio, and some are now troubled by post-polio syndrome, a long-term sequel that occurs decades after infection. Symptoms include fatigue, joint pain, and progressive muscle weakness, afflicting those who long expected their physical problems to be stable. The syndrome may reflect reversal of the recovery processes (nerve regeneration, tissue repair, exercise-induced development) that occurred soon after the original paralysis. Polio sufferers frequently seem to be, or to become, determined, self-disciplined high achievers—people who've had to be much better than ordinary to overcome their disability. Many seem to develop special means of control over their bodies in order to achieve the best that is possible.

Sometimes the spirit of those with paralytic polio is crystallized in a single individual. Franklin D. Roosevelt is the prime example of a man transcending disability, but he is very rarely portrayed in a wheelchair. The poster children of the March of Dimes also make potent symbols of courage and determination. The first, Donald Anderson from Oregon, embodies their virtue: beautiful and brave, he fights his way out of the confines of his invalid cot to stride out toward a future created by the March of Dimes and the people of America.

But another victim of severe paralysis has long captured my imagination and is perhaps the most iconic model in modern America art. She occupies the foreground, a little bit off center, her world extending to a not-too-distant skyline where the brown hillside meets pale cloud. But somehow this horizon is also infinitely far away—a far-flung country on the farthest edge. She is crawling on an unmown field of grass, slowly dragging herself toward a solitary clapboard house eerily unsheltered by its barn and outbuildings. Yet at the same time she is not crawling but remaining effortlessly still, in harmony with a surrounding stillness at whose center she will always be. The scene appears intensely real, as if lit by the perfect light of an American luminist painting—but this is an illusion. Instead, the everyday reality has been stripped away to reveal the mysterious place beneath, a landscape charged with feeling that seems more akin to Dutch paintings of the seventeenth century.

Andrew Wyeth painted this famous picture in 1948. The solitary figure is Wyeth's neighbor, Christina Olsen, who was severely disabled by polio in childhood. Christina was a strong and forceful character without self-pity, and in the painting she exists in a place that seems to lie beyond the strain and struggle that is the daily reality for those who have suffered polio. But perhaps the true meaning of this picture is not expressible in words.

15

GREAT THEMES AND DIRTY LITTLE SECRETS

"One day when the whole family had gone to a circus to see some extraordinary performing apes, I remained alone with my microscope, observing the life in the mobile cells of a transparent starfish larva, when a new thought suddenly flashed across my brain. It struck me that similar cells might serve in the defense of the organism against intruders."

Elie Metchnikoff, 1882

The first 60 years of the twentieth century saw the conquest of polio in America, but the development of vaccines and the science of immunology remained largely separate. Vaccine discovery depended on breakthroughs in bacteriology (as, for example, with the BCG vaccine for tuberculosis), or virology (exemplified by polio vaccines). Meanwhile, immunologists were engaged in the study of antibody: how it worked and how it was formed. All this began to change in 1960 with the discovery of skin graft rejection and a renaissance in the study of cellular aspects of immunity. The notion of the immune system learning the shapes of microbes through exposure to them, so called *instructive* theories, had been abandoned. Instead scientists realized that immune cells naturally recognize the shapes of germs without prior knowledge, and those few cells that recognize the shapes of a particular invader increase their numbers to fight the infection.

But how does the immune system generate the vast array of receptor shapes to fit all possible shapes in the universe of microbes? The answer is that, amazingly, instead of resisting mutation like other DNA, the genes making immune receptors change during the development of an individual because of random rearrangements—a sort of intensive cut and paste of different bits of DNA. New bits of DNA are also added randomly. These genetic recombinations produce many millions of new receptor components, and the random assembly of these components generates still more diversity, creating immune receptors potentially fitting millions upon millions of microbial shapes. The immune system has the potential to recognize or "see" the shapes of all germs before any infection has occurred.

During early development in the human fetus, this vast, new, random family of receptors is screened to weed out those receptors that recognize the *self*—the natural healthy profiles of the body. In other words, immune cells that could attack the normal tissues and organs of the body are eliminated at this early stage. The cells that are left possess receptors with the potential only to recognize invaders. Immune recognition is guided by special structures called MHC (major histocompatibility complex) molecules, which are found on the surfaces of all cells of the body. MHC molecules retain germ fragments in a molecular framework for subsequent presentation to lymphocytes of the immune system. During fetal development the lymphocytes completely unable to bind to the framelike MHC molecules are also destroyed.

There are two major classes of white cells involved in immune defense. The first consists of primitive surveillance cells, bearing special MHC molecules that patrol the body and engulf any invaders they encounter, break them down, and transport germ fragments to their surface MHC molecules. These cells also produce inflammatory signals that attract lymphocytes. It is the lymphocytes that constitute the second class of white cells in immune defense. They are the cells that recognize the shapes of germ fragments.

When lymphocytes arrive at a site of infection, they form intimate links with the surveillance cells in order to inspect their MHC molecules for germ fragments. In this way the surveillance cells "present" germ fragments to lymphocytes. Those lymphocytes whose receptors fit the shape of germ fragments lodged in the MHC molecules then grow quickly, greatly increasing their numbers. The growing army of lymphocytes, all of which have the right-shaped receptors for that particular germ, produce a range of daughter cells. Some give rise to daughters that make antibody, which can neutralize or eliminate the germ. Others recognize infected cells and kill them directly. Still other cells produce chemical messengers that regulate and orchestrate the immune response, and gradually close down the operation when it has been successful. Crucially, when it's all over, both memory cells and antibodies remain to fight off any subsequent infection with the same germ.

The foreign fragment bound in the MHC molecule provides the primary signal to trigger the immune response to that particular germ, but something else is also needed. A crucial second signal is required by lymphocytes or no immune response is made at all. This second signal acts in parallel and is provided to lymphocytes by the surveillance cell presenting the germ. But the surveillance cell only sends this second signal if it has a dangerous pathogen on board. Surveillance cells can sense the general features of dangerous pathogens—viral and bacterial molecules widely present in germs but not found in human cells. These "flags" of the invader constitute alarm signals to the surveillance cell, prompting it to send the all-important second signal to lymphocytes.

During the last decades of the twentieth century, the techniques of molecular biology arrived, and it became easy to produce viral and bacterial proteins in the test tube. Researchers planned to make suitable combinations to produce a synthetic (or subunit) vaccine, with all the right fragments to promote antibody production but without the risks of live infection or killed microbes. However, attempts to provoke immune responses with these purified proteins were mostly unsuccessful. Immunologists had long known of the need for what we call adjuvants (from the Latin *adjuvare,* which means "to help"). These include chemical agents and poorly defined mixes of microbial components—discovered by chance and working mysteriously—which pivotally enhance the immune response.

Adjuvants have always been used by researchers to help provoke immune responses, but they were rarely mentioned outside of the "recipe" section of research reports. No one was certain why they worked, yet they were almost always needed (except with living microbes). In 1989, a prominent and much-admired researcher, Charles Janeway of Yale University, reminded his fellow immunologists of these "dirty little secrets" we all shared yet rarely acknowledged. At that time, immunological theory simply said that anything *non-self,* or foreign (infections, grafts, foreign proteins), triggered an immune response while *self* (our own body components) did not.

But that simply wasn't true. To get a strong immune response, you had to include a bit of witchcraft in the form of adjuvants and related magic molecules. And the more of them the better, as far as immunity was concerned. Janeway proposed that in addition to foreignness, a good immune response needs some indication that the foreign thing is truly *dangerous.* Live infections did this naturally because they had a mysterious something extra. Purified protein fragments of microbes lacked this extra kick. He went on to propose that the ancient first line of surveillance cell defenses, called innate immunity, recognizes molecular motifs widely present on infectious pathogens but not in human tissues.

It is this primitive recognition by surveillance cells (much simpler than the fine discrimination of lymphocyte recognition) that alerts the primitive immune system, prompting it to send alarm signals to the adaptive immune system. This combination of signals then produces a strong and rapid immune response. We now know that this mechanism constitutes a crucial part of the all-important second signal that lymphocytes require before they spring into action. The ancient innate immune system and the "intelligent" adaptive immune system, which recognizes and remembers infections, work closely, hand in hand.

Jenner's smallpox vaccine is a live virus, and although it is administered by an unnatural route, it infects cells and grows in the body, inducing the natural alarm signals and ensuring a good response. A single dose is therefore sufficient to give protection for many years before immunity wanes. We can now add more detail to the picture of what happened when Jenner injected James Phipps: the cowpox virus entered ordinary tissue cells around the injection site and hijacked the normal synthetic machinery of these human cells to make viral proteins and produce more virus. Viral fragments ended up in MHC molecules on the surface of these human tissue cells, marking them out as infected. At the same time specialized surveillance cells of the immune system took up the cowpox virus, digested the proteins, and presented fragments to the small proportion of immune lymphocytes whose receptors could recognize the virus. The surveillance cells, stimulated by danger molecules present in the growing virus, also transmitted the crucial second signal, switching on these virus-specific lymphocytes. These same lymphocytes could also recognize smallpox. The lymphocytes then multiplied; some of their daughters (T-cells) became killer cells, directly eliminating tissue cells carrying the cowpox virus. Other daughters recruited more lymphocytes (B-cells), which also grew and eventually gave rise to cells producing antibodies that neutralized the virus. In this way the growth of cowpox was limited to the local site of injection, though James felt feverish because of the many circulating messenger molecules produced by the immune response. Regulatory immune cells eventually closed down the response when all the cowpox virus had been cleared. Crucially, memory cells and antibody remained long-term. When, seven weeks later, Jenner injected James with living smallpox, the immune system was already primed. Existing antibodies and memory lymphocytes were ready to attack the smallpox virus at once, and the injected virus was rapidly contained. In the long-term, persisting neutralizing antibodies continued to circulate so that James was protected against any subsequent infection with smallpox.

16

THE WAR ON INFLUENZA

"'Ye can call it influenza if ye like,' said Mrs. Machin. 'There was no influenza in my young days. We called a cold a cold.'"

Arnold Bennett, *The Card: A Story of*
Adventure in the Five Towns, 1911

In evolutionary terms, the adaptive immune system and immune memory, on which the success of vaccination depends, began to be added to the primitive first-line surveillance system (innate immunity) about 400 million years ago, when jawed fish appeared. But throughout this time pathogens have also been evolving, and the ancient conflict between infectious agents and the immune system works both ways. Viruses, including smallpox, have evolved many ingenious tricks to evade, confuse, and subvert the immune system. And these viral strategies, like the intelligence service of any great army, include hiding, sabotage, decoys, and impersonation.

Papilloma (wart) viruses hide in the base layer of the skin and do not put on their coat proteins until they reach the safety of the outer layers. Herpes virus hides in nerves and makes periodic forays into the skin, producing cold sores. Cytomegalovirus reduces the number of MHC windows on the cells it infects to evade detection. It also releases protein inhibitors just like those of the immune system, dampening immunity. Smallpox (variola) is among the more ingenious of the viruses: of its 200 proteins, as many as 80 may be agents of evasion. Only two are fully understood: one of them mimics a human switch-off signal for a rapid defense system called *complement. Complement* is the name given to a family of defensive proteins (part of the innate immune system) whose function complements the adaptive immune system. When triggered by danger molecules on microbes, a biochemical cascade of complement proteins mounts an attack on the invaders, punching holes in their membranes. The other smallpox protein that resists the immune response disables an important immune messenger.

Some viruses avoid preexisting immunity by changing their coats. Flu virus does this every year. It happens in two ways: drift and shift. Drift describes small changes in the viral coat that happen all the time. Shift is a sudden, major change, which only happens once in a while. Influenza may infect people, birds, pigs, horses, and other animals, and when cells are infected by two different strains, the genes can combine to form a new virus. Two important flu proteins, H and N, whose job is to gain entry into human cells, protrude like spikes from the virus and exist in different forms. If these appear in a new combination, the human population has no preexisting immunity.

Influenza received its name in 1743. The Italian notion that ills and upsets were caused by the influence of the stars had led to the term *influenza* being used to name outbreaks of disease such as *influenza di febbre scarlattina* (an outbreak of scarlet fever). When an *influenza di catarro,* an outbreak of the catarrhal fever, became an epidemic that spread across Europe, the full phrase soon became abbreviated, and influenza got its name.

In the 1890s, during the golden age of bacteriology when Robert Koch and other scientists were first revealing the true nature of germs, Richard Pfeiffer, who had discovered the antibacterial power of serum from immune animals, isolated a bacterium, *Bacillus influenzae,* from the nasal mucus of a flu patient. It seemed he had discovered the cause of flu. But in 1931, Richard Shope, a young researcher in Princeton, New Jersey, was working on swine flu and showed that it was not bacteria that caused the illness but something that could pass through filters. The cause of pig flu was a virus.

In 1933 Patrick Laidlaw and colleagues at the British National Institute for Medical Research were working on dog distemper, a highly infectious disease caused by a virus related to measles, in a research program funded by pet owners. Dog distemper could be conveniently studies in ferrets, which happened to be susceptible to the dog disease. Laidlaw and his co-workers noticed that when people in the lab went down with flu, the ferrets also started sneezing. By examining ferret nasal secretions that contained the human infectious agent, they showed that human influenza too was caused not by bacteria but by a virus passing through bacterial filters. A young Australian physician was visiting the institute and took an interest in the new infectious agent. When he returned to Melbourne, he managed to grow it in fertilized hens eggs, a technique discovered for other viruses by Ernest Goodpasture in Nashville in the early 1930s. This scientist was Frank McFarlane Burnet, who later turned to immunology and originated the clonal selection theory of immunity, which still stands today.

This work paved the way for flu vaccines. Thomas Francis Jr., who would later become Jonas Salk's mentor and oversee the great field trials of Salk's polio vaccine, was working on flu virus in New York in the late 1930s. In 1941 he was appointed director of the Commission on Influenza set up by U.S. Military Command. The need for a flu vaccine was deemed urgent as war raged in Europe. The military authorities decreed that a flu vaccine composed of killed virus should be developed (though Francis had already worked on live virus immunizations). Flu virus was duly inactivated with formalin, and trials began. The trials revealed the problem of drift and the ever-changing nature of the virus, but late in 1945 mass immunization of U.S. army troops went ahead. Essentially the same vaccine is still used today, with the added strategy of updating the viral strains each year.

The really dangerous complications, and the mortality from flu, are largely confined to the elderly and those already weak or sick or vulnerable because of existing conditions. So why was it that a vaccine for flu was considered imperative by U.S. Military Command on the brink of the Second World War? The answer lies in the momentous events that had taken place a little earlier in the century.

Toward the end of the First World War, in the spring of 1918, soldiers were living in very close quarters in large forts and camps across America, training to join the war in Europe. One of the larger communities was Camp Funston near Fort Riley in northeast Kansas, whose roll included more than 25,000 men. The winter of 1917–18 was a harsh one in the Midwest, and conditions at the camp were not ideal. The region suffered dust storms that blew across the open plain and which the soldiers, and several thousand mules and horses, had to endure. The pall was often thickened by smoke and ashes from bonfires of horse manure that sometimes blotted out the sun. Shortly before breakfast on Monday, March 11, a day or so after a particularly bad dust storm, a company cook named Albert Mitchell reported to the infirmary complaining of a sore throat, fever, and muscle aches. Bed rest was recommended. By noon, 107 soldiers had fallen sick and joined him. By the end of the week, 500 had come down with the illness.

Military camps were places of close confinement coupled with mobility and frequent traffic of personnel—ideal conditions in which infection could spread. The incidence of sickness was also well documented. The Camp Funston outbreak was quickly followed by others in New Jersey, South Carolina, Colorado, and elsewhere. In April and May over 500 prisoners at San Quentin in California were suffering the same condition. This spreading tide of influenza gradually engulfed North America and would eventually become the worst pandemic of contagious disease in modern human history.

How the flu arrived at Funston is not certain, but some experts point to Haskill County in southwestern Kansas, an area from which recruits were drawn. Haskill County was then a rural community where hogs and poultry were raised in simple farmsteads built of prairie sod—ideal conditions for transmission of pig and bird viruses to humans. A local doctor became so concerned about the number of flu cases in his area that in January 1918 he reported the outbreak to a national health agency.

The weight of evidence might favor Kansas as the origin of what was to become the Spanish flu (or La Grippe, or La Pesadilla) that would soon spread across the world, but other possibilities have also been considered. The staging camp for British troops in Étaples-sur-Mer in northern France experienced a mysterious respiratory infection during the winter of 1915–16 that may have presaged the 1918 flu. Étaples was a meeting place for those on their way to the front and those returning battle-weary, many wounded. Conditions there were particularly bad; the English soldier and poet Wilfred Owen, on his way up to the line, described the effects the camp had on morale: "I thought of the very strange look on all the faces in that camp; an incomprehensible look, which a man will never see in England; nor can it be seen in any battle but only in Étaples. It was not despair or terror, it was more terrible than terror, for it was a blindfold look and without expression, like a dead rabbit's."

Just after the Battle of the Somme in the winter of 1916–17, dozens of soldiers at Étaples camp fell ill, complaining of aches, pains, cough, and shortness of breath. Deaths occurred, and some victims had what would later become known as a telltale mark of Spanish flu: their faces became tinged a bluish lavender color, a condition known as heliotrope cyanosis. Two months later a similar outbreak was reported at Aldershot, one of the biggest barracks in England, not far from London.

Whatever the origin of Spanish flu in 1918–19, the first wave of the pandemic was highly contagious. In America, at least, it caused few deaths among the young and fit; it seemed a minor problem in the midst of the Allied war effort as more and more Americans hearkened to the call to arms. In March more than 80,000 U.S. soldiers crossed the Atlantic, and 118,000 followed in April. While sailing across the Atlantic, the Fifteenth U.S. Cavalry reported 36 cases of influenza, resulting in six deaths. By May, the illness was established on two continents.

Both sides in the conflict soon began to suffer the loss of fighting men through flu. In Britain 31,000 influenza cases were notified in June alone, and operations in the Royal Navy were compromised by the sickness of 10,000 seamen. The infection quickly spread throughout Europe and into Spain, where the large number of people affected was widely reported in the absence of press restriction (because Spain was a neutral country). This accounts for the name of the pandemic—elsewhere illness among combat troops was too sensitive to be reported in the press. Despite its rapid spread in the conditions of war, the virulence of the infection remained within the bounds of what was normally expected from flu virus. By early summer, Spanish flu had reached as far as Russia, North Africa, and India; from there it continued its spread to China, Japan, the Philippines, and down to New Zealand. By midsummer of 1918, the flu seemed to abate. Tens of thousands had fallen ill and died, but the worst of it seemed to be over.

Then, late in August, an outbreak centered on the port of Brest in northwest France was carried back across the Atlantic to Boston by returning troops. In September 1918, soldiers at an army base near Boston suddenly began to die. The cause of death was soon identified as flu, but by this time the virus had changed. The second wave was far more lethal than the first. Initially it had been called "the three day fever"—the sickness started with a cough and headache, followed by intense chills and a fever that could quickly rise. It could take a month before survivors felt completely well again.

But the flu that hit New England in the autumn of 1918 was much worse. A doctor at Camp Devens Hospital in Massachusetts described how, two hours after admission, dark brown (mahogany) spots appeared over the cheek bones of the victims; a few hours later a blueness, beginning at the ears, began to spread across the face until it was hard to tell what race the suffering patient came from. This cyanosis was the result of failing oxygenation of the blood in congested lungs. Soon patients would be breathing out a pinkish foam as their lungs filled up, slowly drowning in their own body fluids. By this time they were only hours away from death. This physician estimated 100 deaths each day at Fort Devens in the worst of the epidemic, and he lamented the loss of many nurses and doctors to the infection.

Instead of targeting the old and sick, Spanish flu was particularly deadly among the young and healthy. Almost half of the deaths were among 20- to 40-year-olds, which some experts attribute to the vigorous immune response, in which massive inflammation inundated the lungs. Some military doctors injected severely afflicted patients with blood or blood plasma from people recovered from flu, and this apparently reduced mortality rates by as much as 50 percent. But the number of patients treated in this way was insignificant in the scale of things. Services were often overwhelmed by the number of victims and the swiftness of their deaths. Navy nurse Josie Brown, who served at the Naval Hospital in Great Lakes, Illinois, in 1918, described the morgues packed almost to the ceiling, with bodies stacked one on top of another. The morticians worked day and night. Trucks were used to ferry caskets bound for train stations so bodies could be hurried home. "We didn't have the time to treat them. We didn't take temperatures; we didn't even have time to take blood pressure. We would give them a little hot whisky toddy; that's about all we had time to do." In October alone, 195,000 died of influenza in the United States.

In Europe a civilian population, already battered and bereft, had to deal with the pandemic. London, like other European cities, was ill equipped to cope as troops returned, bringing the virus with them and compounding the tragedy of war. Hospitals were overwhelmed, and doctors and nurses worked to a breaking point, though there was little they could do. Medical schools closed their third- and fourth-year classes so that students could help in the wards. There were no antibiotics to fight the frequent complication of bacterial pneumonia that finished victims off. Penicillin was yet to be discovered. In many towns, theaters, dance halls, churches, and other public-gathering places closed. Streets were sprayed with chemicals and people covered their faces with masks. Some factories relaxed no-smoking rules, believing cigarettes would help prevent infection.

The influenza pandemic of 1918–19 is estimated to have killed 50 million people worldwide, spreading even to the Arctic and remote Pacific islands, making it one of the deadliest natural disasters in human history. Most deaths worldwide occurred in a sixteen-week period, from mid-September to mid-December 1918. In America its impact was so profound as to depress average life expectancy by ten years. Some 500 million (27 percent of the global population) suffered infection.

In 1951 a Swedish American pathologist, Johan Hultin of Iowa University, attempted to isolate the 1918 influenza virus from victims who had been buried in the Alaskan permafrost of a town called Brevig Mission. During the pandemic, 72 of the town's 80 residents had perished from the flu. In his search, he unearthed bodies but failed to detect the virus. Nearly 50 years later, in July 1997, Hultin came across an article in the journal *Science* written by virologist Jeffery Taubenberger of the U.S. Armed Forces Institute of Pathology. Taubenberger had already identified the strain of Spanish flu as H1N1 based on his studies of serum from those who lived through the pandemic. He was now searching for tissue samples of the 1918 flu in order to discover its full structure. Hultin offered his services and returned to Brevig Mission 46 years after his first visit. He was now 73. Local officials, led by Gilbert Tocktoo, Brevig's 35-year-old mayor, agreed that the exhumations would help to gain an understanding of the germ that had killed over three-quarters of their community.

It took five days for Hultin and four helpers to dig through the permafrost, using the long summer days to work from 9 a.m. until almost midnight. At last, between two skeletons, Hultin found the preserved body of an obese woman, roughly 30 years of age. Since her name was not known, Hultin called her "Lucy" (meaning *as of light*) in recognition of the fact that she might throw light on the mysterious virus. The large amount of body fat, and the permafrost, had protected her lungs from decay, and Hultin removed them using a perfect tool—his wife's pruning shears. That evening, in the wood shop of the local high school, Hultin built replicas of the original crosses, and assisted by the students, he restored them to the graves in sight of the sea.

The Brevig material, together with other samples from archived sources, provided scientists with an opportunity to study the actual virus of the great pandemic. The Brevig tissues were chemically inactivated before transport, and one of the four samples contained viable genetic material that Taubenberger was able to exploit. In a series of papers, the team eventually published the complete genome of the 1918 influenza virus, which provided key molecular information as to why the virus had proved so deadly.

Since 1918, a great deal has been discovered about the nature of flu viruses. The two principal types of influenza virus, A and B, are genetically similar enough to be members of the same virus family, but there are a number of differences between them. Type A is more common: it mutates more rapidly (generating several subtypes and many strains) and generally produces more severe illness. Type B virus at any one time constitutes a single strain because it typically mutates only once every few years rather than every year. All the modern global flu pandemics, beginning with the Spanish flu, have been caused by Type A. Bird species are the natural hosts of influenza A viruses, but pigs and other animals also get infected. In contrast, type B infects mammals such as pigs as well as humans, but not birds. The 2009–10 pandemic, the first flu pandemic for 40 years, was caused by the influenza A (H1N1) strain (commonly known as swine flu). Although the pandemic has ended, this strain continues to circulate and cause disease in many countries and was recommended for inclusion in the 2011–12 seasonal flu vaccine.

The World Health Organization has a global network of surveillance and testing that enables it to identify viruses and predict which flu strains are likely to be prevalent each year. As soon as manufacturers get this information, they begin to make a new seasonal flu vaccine with two predicted A strains and the one predicted B strain. The viruses are grown in chicken eggs, where they multiply in chick embryonic cells. The virus is then killed with formaldehyde, purified, and packaged in vials or syringes. Manufacture takes about six months. Flu vaccine manufacture still uses techniques developed in the 1940s. So called *split virion* vaccines consist of the H and N spikes on the viral surface split off from the viral particle.

But the science of flu vaccination is set to change. New vaccines and new vaccination principles are being explored, with some ingenious strategies already in use. One of these employs a living flu virus, which survives and grows only at temperatures less than 25°C. The weakened virus grows well in the nose and throat, stimulating the immune system, but once it reaches the lungs, the warmer temperatures destroy it before it can cause the symptoms of flu. Another route to improving the effectiveness of flu vaccines is to add new adjuvants. Combinations of natural and synthetic molecules that mimic the danger molecules present on natural pathogens are being included in new vaccines to provide alarm signals; these convince the immune system it is dealing not with viral protein fragments but with a dangerous viral invader.

The ultimate goal of influenza vaccine research is the discovery of a universal vaccine for all flu types, ending once and for all the need to reengineer new flu vaccines each year. We are now coming closer to this ultimate aspiration. By screening large numbers of antibody-producing cells from individuals who are making immune responses to flu, researchers have found rare antibodies that neutralize many subtypes of influenza A. These antibodies recognize regions of the H spike on the virus that are hidden away in the viral structure but are shared across subtypes. Such cryptic shared viral structures, normally inaccessible to the immune system, may have the potential to become the universal flu vaccines of the future.

17

FORGED IN THE CRUCIBLE OF WAR

"Seek the truth. The purpose of surveillance systems is to discover the truth."

William Foege, *House on Fire*, 2011

In many ways the first vaccine was also one of the best. Jenner's discovery of a live vaccine that was safe in most people, and which produced immunity for many years with a single dose, remains hard to match. Among modern vaccines we have live attenuated germs such as measles, killed vaccines such as flu and Salk polio vaccine, and chemically inactivated toxoids such as tetanus. The problem with killed vaccines is they're not particularly potent. Killed germs don't infect tissue cells, which means they don't switch on the cellular side of the immune response. With killed flu vaccine, the antibody response is to the viral coat, which changes every year rather than to the core, which doesn't. In contrast, live vaccines are more potent but carry the risk of producing disease in individuals with a weakened immune system, as, for example, with the Sabin polio vaccine.

Because of Jenner's vaccine, the severe form of smallpox (*Variola major*) was no longer a native disease in Britain by the early years of the twentieth century, though it was still imported and produced frequent outbreaks until 1930. After that, outbreaks of imported smallpox fell, except for an increase that occurred after troops returned home from World War II. The milder form of smallpox (*Variola minor*) remained endemic: an outbreak in 1927 infected nearly 15,000 people, but only 47 died. By 1934 it too was no longer a native disease. In North America, severe endemic smallpox disappeared by the late 1920s, though the milder form persisted, infecting 49,000 in 1930. The situation in Europe was broadly similar, though Switzerland suffered an extended epidemic of *Variola minor* in the 1920s. In the 1940s Europe and America finally became free of smallpox.

But severe smallpox remained rife in large parts of Asia. China suffered widespread infection, and vaccination was limited until a national eradication campaign was instituted in 1950. Many Eastern nations were afflicted by *Variola major* in the 1960s. Endemic smallpox was common, too, in sub-Saharan Africa. But by 1970, 170-odd years after Jenner had dreamed of "the annihilation of the smallpox, the most dreadful scourge of the human species," the dream seemed close to becoming reality.

It's difficult to mark the moment that the campaign to rid the world of smallpox really got underway. The fact is, there are several beginnings. One of them took place on February 24, 1947, when Eugene Le Bar, a 47-year-old American merchant and importer, boarded a bus with his wife in Mexico City, bound for New York. Le Bar felt fine when he got on, but that evening he felt sick. The symptoms were mild—a headache and pain in the back of his neck for which he took aspirin. Two days later he developed a rash. Five days later they arrived in New York City, and because of Eugene's sickness, they decided to break their journey. They registered at a midtown hotel. Despite not feeling well, Le Bar and his wife did a bit of sightseeing. They walked along Fifth Avenue and wandered through a grand department store.

On March 5, Le Bar was admitted to Bellevue Hospital, where he stayed for two days. Because of the rash, he was transferred to Willard Parker Hospital for Infectious Diseases in Manhattan. There the diagnosis of smallpox was considered but rejected when doctors failed to find the virus in skin biopsies. The rash, too, wasn't typical of smallpox. Eugene Le Bar died on March 10, yet to be correctly diagnosed. The cause of death was recorded as bronchitis and hemorrhages.

Also in Willard Parker Hospital were Ismael Acosta, a 27-year-old hospital worker from Harlem down with mumps, and an infant girl with tonsillitis. Neither had been vaccinated against smallpox. Not long after she'd recovered and gone home to the Bronx, the little girl developed a rash. She was readmitted on March 21 with "chickenpox." The next day Ismael Acosta, who had been discharged, was admitted to Bellevue and soon transferred back to Willard Parker, also with "chickenpox."

Quite soon a tentative diagnosis of smallpox was made for both patients. This would be confirmed when smallpox virus was grown from their skin samples in chick embryo cultures. Immediately all employees and patients at the Willard Parker Hospital were vaccinated. The case of Eugene Le Bar was reexamined and the truth revealed. All contacts of the three were traced and immediately vaccinated, including guests at the hotel where Le Bar had stayed.

A two-year-old boy who had been suffering from whooping coughing in Willard Parker Hospital during Le Bar's stay was next to come down. Then, on April 6, Acosta's wife, Carmen, age 26, was admitted to Willard Parker. She was carrying their unborn child. It seems she had not been vaccinated. She died of smallpox on April 13.

Another three men, all patients at Bellevue Hospital when Acosta was there, developed smallpox rashes on April 10 to April 13 and were transferred to Willard Parker. The same day, a four-year-old boy who'd had scarlet fever was routinely discharged to a convalescent home in Millbrook, New York. He was to be the source of three other cases at the home: a 62-year-old nun, a five-year-old boy, and a two-year-old girl. Mrs. Le Bar was located by the U.S. Public Health Service in Maine. She had continued on the journey she had planned with her husband. Fortunately she had been vaccinated in earlier life and remained in good health. At this time the outbreak thus consisted of twelve cases, nine originating in New York City and three in Millbrook.

Follow-up along Le Bar's bus route stops in Texas, Missouri, and Pennsylvania revealed no cases of smallpox. An emergency plan for vaccinating all the people in New York City was drafted by health department officials in conjunction with the Commissioner of Hospitals. The Health Department's Bureau of Laboratories was placed on emergency. A statement was issued to the press and messages were broadcast on radio, urging all New Yorkers to be vaccinated without delay. The United States Army and Navy sent several thousand vaccine doses gathered from all parts of the country. Since millions of units were needed to vaccinate the people of New York City, the mayor made an urgent appeal for supplies to support universal vaccination, and manufacturers responded with a 24-hour schedule, packaging their bulk vaccine and diverting all available supplies to the city. Vaccination clinics in all government health centers and all city hospitals stayed open day and night, seven days a week. If things went the way they had done in other U.S. cities in the twentieth century, several thousand New Yorkers would be infected, and almost a thousand would be sure to die. This was *Variola major,* the severe form of the disease.

Rumors ran riot during the vaccination campaign. Many people with chickenpox, together with their families, believed themselves doomed. The health department was flooded with reported cases of smallpox that turned out, on investigation, to be chickenpox. In New York City, thousands of people routinely get sick each week, and about 200 of them die each day, but since the great majority was vaccinated in a very short time, vaccination often got the blame for inevitable illness.

In a period of less than a month, well over 6 million people were vaccinated in New York City, 5 million of them within two weeks of the mayor's appeal for universal vaccination. At the peak, a half-million people were vaccinated in a single day. Never before had so many people been vaccinated in such a short time and at such short notice. Thanks to the health officials and practitioners of New York City; to the press and radio broadcasters; to the American Red Cross; the Women's Voluntary Services; former air raid wardens and other volunteers; and to the co-operation of the citizens of New York, the outbreak was contained. The twelve cases already described were the total number of cases in the New York City smallpox outbreak of 1947.

In 1950 the Pan American Sanitary Organization, prompted both by the dangers highlighted in New York and the extraordinary success of emergency vaccination, adopted a proposal by veteran campaigner Dr. Fred Soper. They began a hemisphere-wide campaign to eradicate smallpox. Soper had become a legendary figure because of his vision and ambition for disease eradication. He persuaded the ministers of health in almost all the countries of the Americas to cooperate in the vaccination campaign against smallpox; simultaneously, he championed regional campaigns to treat yaws (a serious bacterial skin infection producing sores) with penicillin and to eradicate malaria through mosquito control. Despite its limited resources and the broad nature of the drive against disease, the Pan American Health Organization made good progress: by 1957 the chain of smallpox transmission had been interrupted in many Latin American nations. But Brazil, the largest, did not partici-pate in these initiatives, and the impetus was lost. Further resolutions were made, but little progress followed. The World Health Assembly discussed the topic at great length, but their deliberations never crystal-lized into purposeful resolve.

Then in May 1958, as the Cold War deepened and the nuclear arms race gathered pace, Viktor Zhdanov, deputy minister of health of the Soviet Union, addressed the Eleventh World Health Assembly in Min-neapolis, Minnesota. Zhdanov began by tactfully reminding the audi-ence of Thomas Jefferson's letter to Jenner in 1806, which predicted the eradication of smallpox for all time. Describing his visionary plan to rid the world of smallpox, Zhdanov spoke of mass vaccination campaigns but also of the Leicester method—the prompt identification, notifica-tion, and isolation of new cases plus quarantine of contacts, which was held so dear by the fervent antivaccinationists in the city of Leicester 60 years before. It was essential, Zhdanov urged, to use this strategy hand in hand with vaccination. His passion, conviction, and optimism (he boldly claimed it would take ten years) succeeded, and the World Health

Assembly, the decision-making body of the World Health Organization (WHO), approved the principle of global eradication.

At least now there was a common goal, but in practice, many of the countries most affected by smallpox were struggling with malaria and had few resources to divert. Between 1958 and 1966 several nations carried out the program. Some seemed to be successful, but the effort was dogged by enormous difficulties. One of these was wishful thinking in reporting. It was later discovered that some countries reported a nil incidence for purely political reasons, as if smallpox could be defeated by government decree. Another was the focus on mass vaccination, with 80 percent coverage as the target but without the Leicester method of quarantine so essential to success.

Despite the difficulties, there *was* progress: in 1959 around 60 percent of countries had endemic smallpox, and by 1966 this had been halved. But the remaining territories were the truly difficult ones. As well as Brazil, endemic smallpox was still present in most West African countries, in the populous East African nations between the equator and the Tropic of Cancer, in India and Pakistan, and in the archipelago of Indonesia. Some new impetus was needed. At last, in 1966, the assembly heard representations from several delegates for a special dedicated budget in order to revitalize and intensify the stalled campaign.

The intensified WHO global campaign began in February 1966. A dynamic American physician and epidemiologist, Donald Henderson, who had created a U.S. smallpox surveillance unit at the Communicable Disease Center in Atlanta, Georgia (soon to be renamed the Center for Disease Control [CDC]), headed the effort. From his office in Geneva, he considered the global picture the new campaign would have to face.

By this time, smallpox was confined to the tropical regions of the globe. Forty-two countries were reporting cases of smallpox, and 30 of these had endemic disease. A little over 130,000 cases were reported that year, so the task, though difficult, looked as though it might be possible.

But the problems facing Henderson and his team were in fact much greater than they thought. New surveys in Africa and Asia revealed that less than 5 percent of smallpox cases were officially reported. The campaign had, not 130,000, but 2.5 million cases to contend with! The amount of vaccine needed was between 200 and 250 million doses, which more than absorbed the entire program budget of $3 million. Undaunted, the new team set about soliciting donations of vaccine since it was in the interests of every government in the world to eradicate smallpox. The Soviet Union was generous, donating 140 million doses each year for the first few years of the intensified campaign. Forty million doses a year came from the United States. Twenty other countries also donated vaccine, which was tested in two designated laboratories in Canada and the Netherlands.

The vaccine was still essentially Jennerian: calf lymph (or sheep or water buffalo) taken from pustules on the animals' skin, with added glycerin, carboxylic acid, and a digest of proteins to protect the virus. But in the tropics, better preservation was needed. Freeze-drying, a technique already widely used elsewhere, proved ideal. The vaccine was rapidly frozen in the laboratory, and then water was quickly evaporated. Before use it was reconstituted with water and glycerin. Stored as a freeze-dried product, the vaccine was good for months at 45°C. Extreme measures would no longer be needed (e.g, in Peru smallpox vaccine had been carried in kerosene-powered fridges on the backs of mules).

The global program always focused on local resources, but early on there were problems. Local producers made some freeze-dried product containing no virus at all. The potency test was to count how many pocks the vaccine produced on the membrane of a fertile hen's egg. Nevertheless, toward the end of the program, 80 percent of vaccine was produced in the countries using it.

At the outset, four regions of the world were marked out as independent zones unlikely to infect each other: South America, Africa, mainland Asia, and Indonesia. Of these, Africa was considered to be the most challenging continent of all. The stated aim of the global strategy was mass vaccination with rates as high as 80 percent. As a follow-up, vigilance was to be kept high, with rapid containment through strategic vaccination to quell any new outbreaks. But in West Africa, in the first years of the intensified campaign, something new began to emerge.

In Nigeria, the smallpox eradication program had been conveniently combined with a CDC program, backed by the U.S. Agency for International Development (USAID), to control measles, a major cause of death in childhood. Vaccination stations administered both measles vaccine to children under six and smallpox vaccine to everyone. Teams moved from village to village, conducting mass immunizations in which local people lined up to receive their shots. Injection was already viewed favorably in the region because of the great success of antibiotic injections for yaws. Crowd control was important, with roped-off areas where people lined up to make sure things went smoothly. During the day, villagers were often summoned from surrounding areas by tribal drums. With practice, up to 1,000 people an hour could pass through the vaccination station, where an automatic "jet" injector was used to quickly and repeatedly deliver vaccine into the lower layers of the skin by means of high-pressure air. This avoided the need for hypodermic needles.

But because of logistic difficulties and the problem of getting supplies to remote regions, another layer of strategy was tried. One pioneer of the campaign in East Nigeria was a young epidemiologist from the American Midwest named William Foege. At six feet seven, Bill Foege was an imposing figure blessed with a tireless missionary zeal. He realized that tapping into the network of local missionary contacts was an excellent way to pick up intelligence on new smallpox cases. Each evening at seven, the missionaries made contact by shortwave radio to check on their communities, and Foege realized what a valuable resource this might be. He arranged for them to send out runners to their local villages to ask for news of smallpox. The mobile team could then act rapidly, predicting likely movements from known patterns of work and travel, and move in promptly to vaccinate villagers and more distant contacts. In this way the chain of infection could be quickly broken. This microstrategy proved to be extremely successful, not least because, although smallpox is a terrible disease, it is not as readily transmissible as flu or measles.

Despite the surprising degree of success, Bill Foege and his coworkers continued mass vaccination in line with the official plan, reaching high levels of coverage. But quite soon the unexpected happened: in an area of very high coverage, smallpox broke out. It soon emerged that adherents of a particular religious sect had avoided vaccination, and the members were spread out across the region. Once they realized that mass vaccination could fail disastrously, the surveillance-containment strategy seemed all the more attractive. Increasingly, the vulnerability of mass vaccination and the remarkable effectiveness of surveillance-containment occupied their thoughts. Eventually they persuaded the Nigerian eastern regional director, Dr. A. Anezanwu, to let them try their new strategy officially while continuing measles mass vaccination. Independent thinking was very much in keeping with the spirit of the place and time: Anezanwu was an Igbo in the land that would soon declare its independence from Nigeria as the fledgling nation of Biafra.

Operations soon became difficult in the brooding tension of the growing conflict, both on the ground and in emerging policy. The Nigerian government decided to halt vaccine supplies to the east until the rest of the country had caught up with this region.

In March 1967 Bill Foege drove to Lagos for the legitimate purpose of picking up jet-injector parts. He and his CDC colleague began the return journey from Lagos the same day with the parts they required, but they'd also filled their white Dodge pickup truck chock-full of vaccine supplies they'd removed secretly from the secure depot. Taking this dangerous chance was to be crucial in the battle against smallpox as war drew closer.

Foege and his colleague approached Onitsha Bridge at twilight. This was their way back into East Nigeria across the Niger, the great river of

West Africa. They found the bridge blocked with bulldozers and trucks in preparation for impending war. When at last they managed to speak with the commanding officer, they explained the urgent need to get the vaccine to refrigeration at their base in Enugu, the eastern region capital. The officer considered the situation, then ordered a passage to be cleared. The white Dodge continued on its eastward journey, arriving at midnight.

This daring initiative in the buildup to war (deportation was the least of the potential consequences for Foege) was to become an essential link in the chain of activities that led to the adoption of surveillance-containment as the primary strategy for the entire global WHO campaign. The effectiveness of the practice continued to be tested in the field in East Nigeria and repeatedly proved its worth. In July 1967, six months into its deployment, Foege and his colleagues were called to a central CDC campaign review in Accra, Ghana. By this time Biafra had declared independence, and the border between the east and the rest of Nigeria was closed. But the team managed to cross the Niger by canoe at an unofficial crossing point and continued their journey to Accra. Once they reached the Ghanaian capital, they were able to report on the success of their new surveillance-containment campaign.

During their stay in Ghana, the Biafra war began in earnest. In the three years of the conflict, more than a million people would die in battle or from the famine caused by war. With war raging back and forth, Bill Foege was unable to return to East Nigeria. He was eventually arrested because of his links with the eastern region and deported from Kaduna. Not knowing the outcome of the East Nigeria campaign against smallpox, Foege could not return until after Enugu fell to the army of the Nigerian Federal Military Government in October 1968. At last he was able to find out that the surveillance-containment campaign they'd been forced to abandon *had* succeeded: there were no cases of smallpox in East Nigeria throughout the two and a half years of fighting. The Leicester method, transformed beyond recognition by the tropic distances of East Nigeria, had been reborn. And what was to become the primary global strategy to rid the world of smallpox for all time had been forged in the crucible of war-torn Africa.

18

SMALLPOX IN A LAND OF ANCIENT WISDOM

"The examples [of courageous dedication] *are many: an Afghan surveillance team who travelled for four days on foot and horseback through meter-deep snow to reach an outbreak; an Indian District Health Officer and his team who were captured by tribal villagers and saved at the last minute from execution by burning, but who returned to control the outbreak; a Pakistan surveillance agent captured and held for ransom by insurgents, but who on return refused transfer from his assigned area."*

Donald Henderson, 1975

The first of the four World Health Organization autonomous global regions to become smallpox free was South America. The last case was reported in Brazil in April 1971, some nine months after the last smallpox case in West and Central Africa. The second was Indonesia, whose thousand inhabited islands became free of smallpox in January 1972. Eradication was officially confirmed two years later, in each case by an international commission. But in the Indian subcontinent, where mass vaccination was deemed to be the strategy of choice, progress was frustratingly slow. WHO directors in Geneva had learned the lesson of surveillance-containment in West Africa but the authorities in India were much more conservative. They were not convinced of the value of the new strategy and continued to believe that traditional mass vaccination was the only possible solution for their country. Some experts urged a target of 100 percent coverage as the only way forward because of such a high population density, but this was obviously impossible.

Mass vaccination campaigns had achieved considerable success in India in the past—often a vaccinated calf with shaven flanks covered with rows of vaccination sites allowed the vaccinator to travel from door to door, scraping off just the right amount for each injection. By 1900, smallpox deaths had been reduced to around 1,000 per million people; this was halved by 1925. To put this in historical context, in London between 1760 and 1770 annual smallpox deaths had hovered around 3,000 per million people. By 1950 the rate in India was down to 100 deaths per million, but campaigns often missed the inaccessible rural populations and instead revaccinated those within reach.

Among the problems in India were poor reporting of smallpox cases by local communities, low vaccine take-rates due to inefficient vaccination (this means the vaccine did not induce a good immune response: sometimes too little vaccine was delivered, or the vaccine might not penetrate the skin sufficiently). Poor coverage of the total population was another problem. With the new WHO campaign in 1966, alarming new facts emerged. One was the extent of underreporting. In 1967, 83,000 cases were reported in target states where smallpox was most rife, but it eventually became clear that only one in ten new cases was being notified. The real total was nearer 830,000. Around this time two technical refinements were adding impetus to the campaign. In 1968 a forked needle was introduced. With beautiful simplicity, this cheap, reusable device held the right small dose of vaccine as a pen nib holds ink, and it penetrated to just the right depth. The other refinement was more widespread use of freeze-dried vaccine.

In August 1967 Donald Henderson wrote to the Indian government and urged the nation-wide adoption of surveillance-containment methods, which had worked well in the state of Tamil Nadu in the south of India. This provoked a difficult debate on the strategy and also on whether the nationwide campaign should continue at all. Many held that surveillance-containment was only suitable in wealthy, well-vaccinated states like Tamil Nadu. But in 1970 the situation improved, and for the next four years Indian government funding was increased four-fold, mobile squads were introduced, and a ratio of one vaccinator for every 25,000 people was achieved. Reporting methods were streamlined, and the effort became more dynamic with less diversion of resources to record-keeping. By 1973, 1 million vaccine doses had been donated by the Soviet Union, but production was switching toward domestic sources of freeze-dried vaccine, sometimes using water buffalo as donors.

Despite these great improvements, new cases of smallpox went up by 25 percent in 1971 and by 50 percent in 1972. What could be happening? The fact was that the false picture created by underreporting was dissolving to reveal the true situation. By autumn of 1973 surveillance-containment began to take over as the primary strategy to grapple with the sheer scale of the emerging problem. The two states bordering Nepal were the most challenging. One was Uttar Pradesh; the other was Bihar, the land of the tiger and home of ancient wisdom and enlightenment, where Buddha had sat beneath the Bodhi Tree.

The summer of 1973 was even hotter than usual as the eradication campaign began another chapter in the fight against smallpox. Bill Foege, strategist and veteran of the successful Nigerian campaign, was assigned as an adviser in this key region. Fifty-six million people lived in Bihar State alone. The fight was carried forward on the ground by Public Health Centers (PHCs) ultimately accountable, through a hierarchy of control, to central government. In the two states of Uttar Pradesh and Bihar, 1,462 PHCs ministered to 145 million people. Faced with high population density, a strategy of searches sustained continuously for six days was adopted statewide in a blitz to search out smallpox cases and contain them by vaccination. Ideally, for each case found, 20 to 30 surrounding households would be vaccinated together with visitors and other contacts. Schools were included. Pictures of smallpox victims were shown to villagers to help them distinguish the disease from other rashes such as chickenpox. The scheme was launched in earnest on October 15, 1973, and each team searched 20 to 25 villages every day for six days. In practice, performances falling well short of this ideal were still effective in breaking the chain of smallpox infection, highlighting the strength of surveillance-containment.

The first revelation of this renewed campaign was the tenfold increase in cases detected compared with the former system of passive reporting. Families had little incentive to report smallpox in their midst, not least because they feared loved ones would be taken from them in the patient's time of greatest need. The first month of the active campaign searched out 6,000 new cases in Uttar Pradesh and 4,000 in Bihar. Ninety percent of districts had smallpox cases; 2,000 villages were found to be infected, and between 10 and 26 percent of urban areas had smallpox infections.

An outbreak was said to be contained after four weeks had passed without a new case. The six-day intensive searches were carried out approximately once a month, and in Bihar, several thousand new cases were discovered in every search. During March 1974, 600 outbreaks occurred each week. After this, detected outbreaks continued to rise, but so too did containment. Typically, detection was not immediate—the primary case often went undiscovered, and the health team only learned about it as secondary cases began to appear. One of the biggest challenges was the dynamic nature of the populations as people streamed to fairs, festivals, and religious gatherings throughout the long dry season.

Another problem was the inalienable custom of visiting the sick, a fundamental right which also accorded with the traditions of the goddess Shitala Mata. Eventually the containment teams admitted defeat in this area and adopted a policy of posting guards in infected households whose job was to ensure that all visitors were vaccinated before going in to comfort the patient. This strategy was highly pragmatic but also expensive in manpower: 5,000 new cases in Bihar required a workforce of 50,000 to deal with them. Bill Foege describes one tragic incident in which youngsters were attending a memorial party for their small friend who had died of smallpox. All the children of the village were invited. Everyone sat on the terrace outside the house to eat their keer, a sweet rice pudding much loved by children. During the party, the dead child's grandmother, bravely fulfilling her sad duties, cleaned out the sickroom and shook the bedding out of the window, beating the quilt to clean and air it. Ten days later all the unvaccinated children who'd come to remember their friend developed smallpox.

Smallpox outbreaks peaked in the April–May six-day search of 1974. At that time evaluation teams were working to understand what went on during search and containment so that practices could be continually improved. In the first four months of 1974, India reported more than 67,000 cases of smallpox, two-thirds of them in Bihar. The land of ancient wisdom on both sides of the Ganges would be the key to success against smallpox in modern India.

April was the tipping point for those, like William Foege, who understood the figures. The number of cases was continuing to climb, the ancient scourge was striking down 1,000 victims every day, but increasingly successful containment was set to outstrip new cases in Uttar Pradesh in May. In Bihar 500 outbreaks were being contained every week.

Bihar that May was extremely hot. The rains would not bring their blessed relief until the end of June, making it a peak time for transmission. The searches continued relentlessly. The sixth search of the campaign revealed 2,622 new cases, which was partly a sign of increasingly effective methods. However, increasingly efficient containment procedures were now clearing outbreaks at the rate of more than 800 a week, 300 more than the previous month. Field workers were stretched to the limit but they were prevailing, and this gave energy to the great endeavor—a public health operation on a scale never seen before.

But suddenly everything changed. A strike on the railways, the largest railway system in the world, took hold and impacted other industries. Soon other workers followed the example, including half the vaccinators in Bihar in a protest against low pay. District medical officers followed suit, planning to walk out by June if their demands were not met. Meetings and negotiations followed, but the great endeavor was in jeopardy.

Then another disaster struck. On a windless Saturday in May 1974, the day of Buddha Jayanti in India—a festival marking the birth of Gautama Buddha—the Indian authorities carried through the project code-named *Smiling Buddha* and detonated their first nuclear device in the great Thar Desert in the state of Rajasthan, the land of kings west of Uttar Pradesh. In response, the United States imposed economic sanctions on India, cutting off all assistance except humanitarian aid, which fortunately included the U.S. contribution to the smallpox eradication campaign. Other nations such as Russia and China refrained from criticizing India. In the wake of President Nixon's visit to China in 1972, tensions between the superpowers were now three-sided. Hostility had lessened, but the Cold War continued and would worsen by the decade's end. To cover the global story, the world press descended on northern India.

This might not have mattered had the press departed when the nuclear story was exhausted, but some elements then turned their attention to the smallpox campaign. This was a politically sensitive area for India. In retrospect it seems clear that the most intensive months of the campaign against smallpox coincided with a smallpox epidemic, which would have limited the progress of even the best-organized campaign. Nevertheless, under the intense scrutiny of international attention, the campaign was criticized and reexamined at the highest level in its most critical phase. While the well-informed predictions of success were all about the turning tide, the number of outbreaks was still climbing (the numbers for May were not yet in), and some highly influential figures believed that a return to traditional mass vaccination was the only way forward for India. The reawakened debate swung back and forth and came to head in late May in Bihar at a meeting attended by the state minister of health.

In the run-up to the meeting, influential officials dedicated to the campaign, including Bill Foege, met with the minister to convince him of the imminent success of surveillance-containment. Their own figures had convinced them they were on the brink of success, but on paper, they couldn't deny that outbreaks were still rising (though clearance rates were catching up). Their private attempts to convince the minister and win a reprieve for the strategy, if only for a further month, were doomed to failure. The minister was concerned that the backlog of unimmunized children in the population was building up as efforts focused exclusively on outbreaks, and he was set on a return to mass vaccination.

The official meeting took place in Patna in the sweltering conditions of the high summer season on Monday, May 27. The room was packed with campaign workers listening to the usual reports and updates as the ceiling fans turned slowly, barely stirring heavy, moisture-laden air. Suddenly, surrounded by his entourage, the minister swept in and proceeded to address the assembly. He spoke of the several thousand active outbreaks in Bihar not yet contained. He lamented the still-rising numbers of new outbreaks outstripping the number of completed containments. It was clear, he told the meeting, that the battle was being lost; he therefore intended to revert immediately to the traditional strategy of mass vaccination.

An anxious silence settled on the crowded room. All the efforts, hopes, and dreams of the last seven months' grueling and exhausting endeavor seemed poised upon that moment.

It was in this seemingly interminable silence that a young Indian physician respectfully raised his hand to ask a question. As Bill Foege would later describe it, he was in no way an imposing or important-looking figure, but merely one among the many dedicated workers in the field. And yet he was about to conjure one of the most unexpected and uniquely compelling images of smallpox and the global fight to end it for all time.

The questioner, spare and slightly built, was visibly shaking as he prepared to say his piece. It was clear he understood what was at stake. He described himself as just a poor village man, but he went on to speak about the certainties of village life. If someone's house caught fire, other villagers ran to help. And they didn't waste time pouring water on all the other houses in the village just in case the fire were to spread, they rushed to pour water where it would do the most good: on the flames of the house ablaze.

It wasn't the burning heat that pressed on every head in the meeting chamber. And it wasn't the thought of precious water in the long, dry summer season yet to end. It was the startling encapsulation of the containment campaign against smallpox in this single brilliant image that galvanized the room.

Another pensive silence fell. All eyes were on the minister. Eventually he spoke. He spoke of the great personal and political pressure he was under. He spoke of the grave risk he was about to take. But, yes, he had reconsidered. And, yes, he *had* changed his mind. He would give the surveillance-containment strategy one more month to prove its worth.

The railway strike was settled in the same week, and other key workers who had threatened action were also mollified. The May smallpox figures in Bihar were at last revealed for the searches conducted in the first week of June. They showed the outbreaks waning by one-third! And the number of individual cases was down dramatically, from 14,000 in April/May to 7,500 in early June. That one precious extra month granted by the minister was enough to reveal the success of the surveillance-containment strategy for everyone to see.

And at last the rains came, bringing their blessed relief, welcomed with rejoicing, greening the Gangetic Plain. The monsoon, as always, further dampened the transmission of smallpox as more people stayed at home.

The same watershed was reached soon afterward in Uttar Pradesh. Cases peaked in June, with almost 800 new outbreaks per week, but then fell quickly. During the next four months, smallpox outbreaks fell dramatically, and by November the number of outbreaks being contained in the whole of India had fallen below 1,000. Throughout this four-month period, no one took a breather and no one slackened pace. Instead the army of field workers redoubled their efforts. There were reversals here and there in smaller states not yet cleared of smallpox. Criticism of the campaign resurfaced now and then. Importation from Bangladesh, where smallpox remained endemic, was sometimes a problem. But the tide was truly turning. The map of outbreaks in Bihar in early 1974, each one marked with a black spot, looks like a rampant smallpox rash. But one year later, in January 1975, there is hardly a pimple to be seen. One crucial contribution to success had been the contribution of an Indian giant, the Tata Iron and Steel Corporation. WHO staff with canny powers of persuasion got the company to take on the campaign strategy in southern Bihar for all their dependents and associated workers. The company tackled this with industrial efficiency.

With the number of active outbreaks in the whole of India below 1,000, it looked as if the end might be in sight as long as nothing dreadful happened in the coming dry season. At this point the ultimate weapon was deployed. Rewards were offered for reporting new smallpox cases. At first the reward was ten rupees (about $1.50). Almost every case had several hopeful claimants. As smallpox got more rare, the reward grew. Toward the end it would reach a staggering $1,000. As long as nothing unexpected happened, the wealth of experience already gained in trends and patterns of smallpox outbreaks enabled campaigners to predict an end to smallpox in India by May 1975.

On May 1, 1975, Campaign Director Donald Henderson was not presiding in Geneva, nor was he visiting workers in the great Gangetic Plain; he was keeping an appointment in the city of Bristol, less than twenty miles from Jenner's house in rural Gloucestershire. Henderson was giving the first Jenner Memorial Lecture at the Bristol Royal Infirmary, and his address was very timely—the campaign in India would soon report its last case of smallpox. The reward for reporting new cases had yet to skyrocket and still hovered at 100 rupees. Henderson spoke of the intensified campaign, the thousands of false alarms that turned out to be chickenpox, the house-to-house searches that were eventually conducted when 100,000 workers visited more than a million households across India. In remote regions of Bihar where ancient tribal cultures flourished, he described the fierce resistance to vaccination, which was overcome by dawn raids on slumbering villages with added police support. Any later arrival could provoke a fusillade of spears. Once vaccination was in progress, the villagers' moods would change dramatically, and the team could sit down to a cordial breakfast with the elders. It was their cultural duty, the elders explained, to do their best to resist vaccination, but once they had no choice, tradition demanded that every hospitality be extended to the visitors.

Henderson spoke of dedicated workers traveling on camels, mules, bullock carts, sometimes on elephants, cycling dirt roads, trudging endless muddy tracks, fording rivers in the glare of summer to search out and contain the invisible enemy. He spoke of the changes made in the WHO to inspire leadership and how, soon after his arrival, several desk-bound bureaucrats had been highly ruffled at the thought of actually venturing into the field. He confidently predicted that the last case of smallpox in India would occur in days, or weeks at most.

And Henderson was right. A 30-year-old woman, infected in Bangladesh, became ill in Assam State on May 24, 1975. She would be the last case of smallpox in India. Her name was Saiban Bibi, a homeless Bangladeshi beggar who lived on the railway station in the small town of Karaminganj. After five days of fever, lying trembling on the platform or in the third-class waiting room, she was taken by concerned passengers to Karaminganj Civil Hospital, where smallpox was diagnosed. This, though they didn't know it, would be the last core task for the containment teams who by this point were truly on top of their game. Twenty-seven thousand people were rapidly vaccinated in Karaminganj alone. Happily, Saiban recovered and no one else became infected.

Toward the end of his lecture in Bristol, Donald Henderson spoke of the kindly country practitioner who had lived his life not very far away and had set in motion the whole immense endeavor to rid the world of smallpox forever. In India, the second-most populous nation on earth, Jenner's dream was becoming a reality.

19

THE FINAL DEFEAT
OF SMALLPOX

"Work in the area of the final inch is very very complex but also especially valuable because it is executed by the most perfected means."

Alexander Solzhenitsyn, *The First Circle*, 1968

As early as August 1970, smallpox had been banished from East Pakistan, now called Bangladesh. This was the first Asian country to be freed of the disease by the intensified smallpox eradication campaign. But the Bengalis who peopled its rich, alluvial plains were thirsting after nationhood. In December 1971 Indian troops and the Bengali rebels of East Pakistan defeated the forces of West Pakistan. The sovereign nation of Bangladesh, led by Sheikh Mujibur Rahman, was born. Early in the new year, vast numbers of Bengali refugees set off for their newly liberated homeland. Many of them were from a camp called Salt Lake, near Calcutta. Unlike other refugee camps, the authorities in the Indian state of West Bengal did not allow WHO or national health staff to visit Salt Lake Camp, and the regime, run by volunteers, had neglected vaccination. Cases of a febrile sickness with skin rash in November were diagnosed as chickenpox. In fact it was smallpox. William Foege, already back home in Atlanta, watched a TV documentary about the camp and its hospital and to his horror recognized the smallpox rash. A chain of telephone calls eventually alerted the authorities but it was too late—50,000 refugees had already begun their journey homeward. Amid the celebrations and the cries of "Joy Bangla!," thousands of returning migrants carried smallpox into Bangladesh. The dreaded speckled revenant regained its hold and spread rapidly across the war-torn country. Leaders of the intensive campaign had no choice but to begin all over again.

By the spring of 1974, victory seemed once more in sight. Bangladesh is a land of rivers straddling the Tropic of Cancer. Its great delta plain at the confluence of the Ganges, Brahmaputra, and Meghna rivers is richly fertile but highly vulnerable to floods. In April heavy rains began and didn't stop. Along the Brahmaputra, huge floods were soon devastating the fields on either side, ruining the rice crop and bringing famine to the land. A great triangle of countryside between Syhlet, Rajshahi, and Barisal, with Dhaka at its center, was eventually inundated, including the northern areas where smallpox was rife. Large numbers of health workers were diverted to flood relief, food distribution, and cholera control, while hundreds of thousands of refugees went on the move in search of food. Smallpox was spreading yet again—one among many ills afflicting the fledgling nation.

In January 1975 the campaign took a third horrendous blow. Almost overnight, 400,000 people lost their homes as the illegal slums of Dhaka were razed to the ground by bulldozers. These, the poorest among people, began to spread across the country, having no choice but to try to return to their homelands and carrying smallpox with them.

In February 1975 an emergency program was authorized by Sheikh Mujibur Rahman. More than 60 international staff were hurriedly brought in. It took two months to bring the explosion of smallpox cases under control. On August 15, Sheikh Mujib was assassinated. In November, Bangladesh reported its last case of smallpox. The victim was Rahima Banu, a three-year-old girl who lived on Dakhin Shahbazpur, or Bhola Island, the largest island of Bangladesh in the mouth of the Meghna River. Rahima made a good recovery and would be the last case of naturally occurring *Variola major,* the severe form of smallpox, throughout the world. (I can't help wondering about the fortunes of Rahima. In 1995, half of Bhola Island became permanently flooded, leaving 500,000 people homeless. The Bhola Islanders have been described as the world's first climate-change refugees.)

The great achievement of eradicating smallpox in a land beset by flood, famine, and disaster was made possible by the efforts of a large influx of WHO international staff. Two hundred experts from 28 countries eventually joined forces with the national services of nineteen civil surgeons at district level, 400 officers at subdistrict (or *thana*) level, and 10,000 family welfare workers at village level, all aided by the resources of government health agencies, international charities, and volunteer organizations around the globe.

Meanwhile, smallpox persisted in the eastern region of Africa and was endemic in the kingdom of Ethiopia. This ancient land is bound up with the story of Eden, where, according to the King James Bible, four rivers watered the Garden, "and the name of the second river is Gihon: the same is it that compasseth the whole land of Ethiopia." The highlands, which cover most of the country, provide a cool climate compared with other equatorial regions. Warm days and cool nights are largely free of humidity, and some liken it to an eternal spring. Most of Ethiopia's major cities are high, including Addis Ababa and the ancient capitals of Gondar and Axum. The vast highland complex of mountains and dissected plateaus is divided by the Great Rift Valley, which runs southwest to northeast and is almost certainly the cradle of the human species: the bones of hominids from several million years ago have repeatedly been unearthed. But Ethiopia is also a diverse country of deserts, tropical forests, lakes, and rivers, including the once mysterious source of the Blue Nile. This great diversity means wide variations in climate, soils, vegetation, and, most of all, in human settlement to the remotest reaches. Some 90 languages are spoken. More than half the people live more than a day's journey from any road. In the central and western highlands, communities are small, and the landscape, dramatically etched with deep ravines, is covered with forest. Tackling smallpox in this country would be a challenge indeed.

Here the endemic kind of smallpox was *Variola minor,* the less aggressive viral strain, which meant that while mortality was low (around 2 percent), vaccination against a mild disease seemed relatively unimportant and reporting was not a high priority among native peoples. In the autumn of 1973, as the campaign was succeeding in India, a severe drought in east and northeast Ethiopia led to crop failure. Tens of thousands starved to death, and in the flight from famine, smallpox was carried south and west across the country and over the border to Somalia. In response to the emergency, France, the Sudan, and Kenya provided aid, and the U.S. Public Health Service financed three Canadian helicopters for use in the campaign against smallpox.

Further turbulence lay ahead. Early in 1974, while the campaign worked to contain smallpox, an army mutiny sparked by shortages of food and water quickly spread to other groups in a land made restless by high fuel prices, lack of schools, low pay, and poor conditions. Rebel groups demanded a new political system. Eventually, Emperor Haile Selassie was deposed and a communist military junta came to power.

Despite the revolution, which led to chronic conflict along Ethiopia's borders, smallpox cases fell; many regions became free of disease; only the mountains of the central zone still harbored the infection. During the turmoil, one of the campaign's helicopters was shot down, and another was destroyed in a grenade attack by villagers who believed their Italian occupiers of 30 years before were returning. No pilots were killed, but one was held for ransom for a time, during which he succeeded in vaccinating his captors. He was later released for a modest sum and resumed his duties. The new year of 1975 saw many areas cut off by civil war, but the great success in Asia had freed up resources, and the global campaign began to close in on disease in Ethiopia, the last place in the world afflicted with smallpox. The revolutionary government gave smallpox eradication a high priority. More helicopters were provided along with complex logistical support essential in the difficult terrain. The intensified campaign focused on the central highlands, the vast Blue Nile Gorge, and the Ogaden Desert in the south, whose nomadic peoples possess a heritage not bound to any modern state.

The central highlands were the first regions to be freed of smallpox in July 1976. Later in the same month, an outbreak was detected in a nomad village called Dimo in the southern Ogaden, an arid country of dense shrub land, grasses, and bare hills. In fact, the outbreak had been in progress for some time, but conflicts between Ethiopian forces and guerrilla fighters made the area hard to monitor. By the time it came to light, there were sixteen cases; the last to resolve was a three-year-old girl named Amina Salat.

Some of the great themes in the story of smallpox have barely changed through history. In Africa in 1976, the age-old custom of variolation, introduced to Europe's aristocracy by Lady Mary Wortley Montagu in 1721, was still practiced. It was still common among certain ethnic groups, including nomads of the Ogaden. As many as 12 percent of smallpox cases could be attributed to variolation. Amina Salat had been variolated and was showing signs of the disease caused by the procedure. The wisdoms of an ancient tribal people had collided with the modern age. The vaccination team reached her in the early stages but was too late to suppress the symptoms, though she recovered well. For a time it seemed as though Amina might be the last case of smallpox in the world. By September some officials were beginning to think about a celebratory press conference. But then news came of an outbreak in Mogadishu. The disease had spread to Somalia.

Some reassurance was taken from the fact that the Somali cases, five in all, had traveled from the Ogaden—the disease seemed to be imported. But by October the number afflicted had risen to twenty, and by year's end it was 34. Gradually the true picture began to emerge—smallpox outbreaks had been hushed up in Somalia to avoid official embarrassment. An unofficial register of hospital admissions would later show more than 500 smallpox cases by late 1976.

In 1977 the Somali campaign appointed a new manager, and an intensive search was undertaken. In March, following the successful

Indian strategy, a reward was offered—200 shillings (U.S. $32) for each new case reported. Two cases soon appeared in southern Somalia near the Ethiopian border, and by April outbreaks in different locations had climbed to 40. The disastrous resurgence continued in May, when 600 cases were reported. Kenya and Ethiopia were under threat of re-infection, and if things got worse before the *hajj* in autumn, the disease could be carried out of Africa to Mecca and the greatest gathering of pilgrims in all Asia.

At this point the Somali government declared a national disaster and appealed for UN help. Funds and material aid of all kinds began to flood in, and not a moment too soon. A highly experienced WHO adviser, Zdenck Jezek, was appointed to head the campaign, and he arrived in Mogadishu on May 10 as the outbreaks continued to climb. His first act was to strengthen operations by increasing team numbers and introducing interweaving paths of surveillance to ensure contact with elusive nomads journeying through regions of high scrub.

It is perhaps not surprising that the last refuge of smallpox was an ancient wandering people, often persecuted, often shunning contact with the modern world. It was especially difficult to reach them as their homeland lay beneath the shadow of armed conflict between Somali and Ethiopian fighters. Once located, the search for vaccination marks among these people was often hampered by traditional skin decorations. Special isolation camps were established, but patients often left while still infectious. Later, small enclosures with thorn-bush fences inside nomad encampments proved helpful, with patients kept close to their families. Quarantine was sometimes rewarded with payments and new clothes so contaminated garments could be burned.

The smallpox epidemic peaked in June when new cases reached more than a thousand. By this time, southern Somalia had become the focus for the entire global campaign. A force of 3,000 local workers was rapidly trained to deal with the emergency. Vaccination levels in Somalia reached 70 percent in August and climbed to 90 percent in September. At this point the number of cases began to fall dramatically.

In October 1977, as smallpox dwindled to nothing under the onslaught of intensive containment, a health officer drove a mother with her two sick children from an outlying village into the port of Merca, 45 miles southwest of Mogadishu. Sadly one of these children would later die of her disease.

The health officer trying to help this family was searching for the isolation hospital and, not knowing where it was, asked for directions at the general hospital. He was pleasantly surprised when the young man of whom he'd asked the way climbed into the Land Rover and offered to guide him there.

This kind-hearted young man was Ali Maow Maalin, a 23-year-old Somali who worked as a cook at the hospital. Ali had done some work as a vaccinator during the campaign and had also received a smallpox vaccination. Unfortunately, it had not been done properly. It was only a short journey—Ali was in the car for just a few minutes. Nine days later he began to feel feverish and had to go home from work. Friends and neighbors called in to see how he was. Three days later he was admitted to hospital with what was thought to be malaria. A skin rash with pustules developed and was diagnosed as chickenpox.

It was only later, after Ali had been discharged and sent home, that a nurse visiting him from the hospital diagnosed Ali with smallpox. A WHO epidemiologist quickly confirmed the case. A large-scale, organized frenzy quickly followed: more than 50,000 vaccinations were accomplished within two weeks in the town and its surroundings. Merca was thoroughly scoured every week for six weeks, and the area around it intensively searched every month for five months. No more cases of smallpox would be discovered and Ali Maow Maalin would be the last case of natural smallpox to be recorded in the world.

20

INVISIBLE WEAPONS
OF WAR

"After the collapse of the Soviet Union, in early 1992, Russian President Boris Yeltsin signed a decree banning all biological weapons-related activity. Considerable downsizing in this area did indeed occur, and included destruction of existing biological weapons stockpiles. However, there still remains doubt that Russia has completely dismantled the old Soviet program."

Ken Alibek (formerly Kanatjan Alibekov,
first deputy director of the Soviet biowarfare
agency Biopreparat, 1988–1992), 1998

Unlike other monumental achievements in human history, the global eradication of smallpox had no sweet moment of triumph to possess and celebrate. There was no single step that represented a giant leap for mankind. It simply wasn't possible to mark the instant of success, because no one could be sure it had happened until long afterward. It took a little over two years for the global commission, set up to certify final eradication, to reach their decision and agree sufficient time had passed. On December 9, 1979, they made their definitive announcement; at the time of this writing more than 30 years later, there have been no more cases of natural smallpox infection.

Is the world now truly rid of smallpox? The answer must be: not entirely. At the annual meeting of the governing body of the World Health Organization in Geneva, Switzerland, on May 24, 2011, a committee of world leaders decided to keep, for the time being, the world's two remaining stocks of the smallpox virus. "We've beaten smallpox once, but we must be ready and prepared to beat it again, if necessary," was the informed opinion of U.S. President Barack Obama's administration. Iran opposed the decision to retain stocks, apparently objecting to a U.S. clause demanding all countries affirm to the WHO that they do not hold undeclared stocks of the virus. The Russians agreed with the Americans—the two remaining viral stocks are held securely at Atlanta, Georgia, in the United States and in Koltsovo in Russia, a purpose-built settlement for the employees of the State Research Center of Virology and Biotechnology.

In every country in the world, the eradication of smallpox has been hailed as a historic success, but in a world free of natural smallpox, human immunity has waned. Routine vaccination ceased early in the 1970s, which means (given the limited duration of vaccine protection) that the general population is as vulnerable to smallpox as were the native peoples of the Americas in the seventeenth century, when entire nations were wiped out. This makes smallpox ominously attractive as a weapon of biological warfare.

In 1969 President Nixon declared that the United States had renounced the use of biological weapons. Research and development at Fort Detrick in Maryland, once the center of U.S. bioweapons research, switched to defense strategies. (The United States Army Medical Research Institute of Infectious Diseases is now a leader in bioweapons defense.) Three years after Nixon's declaration, the Russians agreed. It was a time of detente or *razryadka,* and a thaw was underway in the Cold War between East and West. Despite their declaration, the Russians kept up their production of smallpox virus in a secret facility 40 miles northeast of Moscow at Zagorsk (now called Sergiyev Posad, the site of a great medieval monastery). Here a particularly virulent strain of variola virus, brought by an Indian traveler who infected many Muscovites in 1959, was propagated in hens' eggs. In 1970 the stockpile of stored virus reached twenty tons.

A state-owned pharmaceutical company, Biopreparat, was established as a cover for producing biological weapons. When, in the late seventies, relations worsened and the Cold War intensified, many thousands of workers were employed. Propaganda and paranoia have always driven superpower tensions, and this was particularly true of bioweaponry. Anthrax was another deadly agent once favored by both sides; its development caused a major accident in Russia in 1979 when more than 60 people died. No doubt an atmosphere of subterfuge, intrigue, and coercion prevailed in these top-secret, state-sponsored activities; even so, it is surprising that the man appointed as chairman of Biopreparat was none other than Viktor Zhdanov, the highly principled champion of the modern smallpox eradication movement at the WHO in 1959.

After two years Zhdanov was made to resign, and, probably with great relief, he returned to honorable research in virology while the bioweapons program rolled on without him. In 1980, just as the WHO felt able to recommend an end to unnecessary smallpox vaccination, the cold war deepened and the Soviet Union launched a program to produce still more lethal strains of smallpox and other deadly bioweapon candidates.

In the following year, official stocks of smallpox in the world were reduced to just five: the United States, the Soviet Union, Britain, China, and South Africa. It was highly desirable to reduce smallpox holdings even though these were legitimately used to produce harmless DNA sequences for future research. In 1978, a tragedy had unfolded in Birmingham, England, where smallpox was held at the University Medical School. A 40-year-old technician in the anatomy department became ill while she was working in a darkroom above a laboratory conducting research on live smallpox. The deadly virus probably arrived through a service duct, and she became ill in mid-August. She had been vaccinated against smallpox in 1966, but tragically, on September 11, she died. Many people came into close contact with the smallpox victim before her infection was diagnosed, but only her mother contracted it. She survived the infection. Yet sadly the accident precipitated two more deaths: the woman's father died of a heart attack while visiting his sick daughter, and in the interval between the accidental infection and the laboratory worker's death, the head of the department responsible for the virus took his own life.

By the early 1980s a library of harmless viral fragments was available to research institutes around the world. British stocks were transferred to Atlanta, Chinese stocks were officially destroyed, and so were the South African materials, leaving only the United States and the Soviet Union as custodians. On the face of it, the good sense and high ideals of the eradication campaign had prevailed in international politics. Sadly this was an illusion. What happened next is almost beyond belief.

In the early 1950s the two things most feared by many Americans were said to be polio and a nuclear holocaust, but even the worst paranoia would not have predicted what happened in Russia in the mid 1980s. In 1987, ten years after the last natural case of smallpox, in a Soviet Union where Mikhail Gorbachev was launching his great reforms and ushering in the age of *glasnost* and *perestroika,* Biopreparat was working on a still more deadly weapon for the delivery of aerosol smallpox by cluster bombs. Simultaneously, the Koltsovo production line was being developed to produce enough virulent material to arm these bombs. Research was also going on to genetically modify the variola virus to make it resistant to ionizing radiation—the ultimate completion weapon to follow up a nuclear attack. Plans included delivery by inter-continental missiles and splicing smallpox with other deadly diseases such as viral encephalitis. The deadly Soviet war machine was lumbering onward despite the changes that were beginning to transform the political map of the world.

In 1992, following the collapse of the Soviet Union, Boris Yeltsin at last banned offensive biological research, and the chief scientist at Bio-preparat, Kanatjan Alibekov, defected to the West and told his story. He also warned that although most of the weapons capabilities had been de-commissioned, a diaspora of disaffected Biopreparat scientists had taken place, and some individuals may have carried vials of smallpox to countries across the world, including rogue states where they could potentially be used as weapons. In light of this, it is perhaps not surprising that the U.S. and Russian governments remain reluctant to destroy the last official smallpox stocks; in all likelihood they're not really the last stocks in the world. I had the privilege of visiting the research establishment at Fort Detrick, Maryland, in the summer of 2000 and was impressed by their high-powered research on vaccines against weaponizable micro-organisms and their continuing determination to protect troops in any future conflict.

In 2003 Donald Henderson, whose responsibilities by then included chairing a U.S. advisory council on the threat of bioweapons, warned a conference in Geneva that Iraq, Syria and Iran were amongst the states that might have retained stocks of smallpox after the epidemics of 1970–72 brought by pilgrims from Afghanistan. Soon afterwards the CIA issued a report: *The Darker Bioweapons Future,* warning of the increased risk brought by advances in genetic engineering which could be used to produce infections of a virulence beyond that seen in nature.

Of course, such reports focus on a worst-case scenario, but one effect is to ensure that funding continues for research on smallpox, including medicines to treat it that could also find broader uses. There has never been a successful medicine to treat smallpox, but this may change in the future. Researchers at Fort Detrick found that a new antiviral drug for herpes could also be used against smallpox, and improvements to the drug's safety profile have been made. Other researchers have focused on the mechanisms by which smallpox evades immune defense. They're working on ways to disable the viral proteins that hoodwink the immune system.

The current strategy adopted by several governments to protect their people centers on readiness to vaccinate in the event of an emergency. Routine vaccination is no longer justified, even for special groups, because of the damage the vaccine can cause in a small minority of people.

It would be much easier to apply this strategy if a safer vaccine were available. The genetic engineering that can make smallpox even more virulent can also be used to make a vaccine virus safer than the one used in the eradication campaign. These attenuated vaccine viruses can also be modified to carry fragments of other viruses. In this way they become potential vaccines against many other diseases. Modern laboratory studies of this kind revealed, quite early on, an amazing fact about the successful smallpox vaccine that had been propagated in cows and water buffalo in so many nations of the world. The vaccine used in the eradication campaign is *not* cowpox.

This extraordinary discovery throws into relief the mystery surrounding the invisible universe of germs that prevailed for so many centuries. At first the cowpox infectious agent was defined as a poisonous principle manifest as pustules, described by one practitioner as glistening with a luster of silver or pearl. This "virus," or poison, produced consistent patterns of symptoms distinct from similar infections such as pseudo cowpox, and Jenner understood this well. Later, cowpox was conceived to be a live infectious agent that could pass through the finest filters. Later still it was characterized as a particle of life visible under powerful microscopes. And in the age of molecular biology, it was understood to be made of DNA encoding proteins that hijack the synthetic machinery of cells to reproduce the virus.

One thing this complex level of description indisputably reveals is that cowpox and the modern vaccine virus, now called *vaccinia,* are closely related but are nevertheless clearly different species. It seems likely that for much of the nineteenth century, the Jennerian vaccine *was* cowpox; but at some point, or at several points in countless transfers and propagations of the virus, vaccinia replaced cowpox. All strains of vaccine used in the twentieth century appear to have been vaccinia. So where did this virus come from?

Vaccinia is in fact a collection of virus strains maintained in different research and production laboratories over the decades, all variants of a single species belonging, like smallpox and cowpox, to the poxvirus family. To this day a mystery surrounds its origins, and in its modern form it doesn't naturally infect animals. As Derrick Baxby, an authority on both viruses and Edward Jenner, has pointed out, its true origins will probably never be known. The most likely theory is that it came from horsepox. Horses infected with horsepox (not the infection called grease that Jenner spoke about) were used alongside cowpox as a source of Jennerian vaccine, and at some point a horse-derived poxvirus may have taken over from cowpox in the propagation of vaccine stock. Unfortunately, horsepox is now extinct so no molecular comparisons are now possible.

But it all began with cowpox, and we now know a little more about this rare infection. It seems that cowpox, despite its name, is not naturally maintained in cattle but infects them accidentally. The virus is maintained naturally in small rodents such as bank voles and wood mice across Europe. Apart from cattle, who share their pastures with mice and voles, another obvious accidental victim of cowpox virus is the domestic cat—"the little tyger of our island" as Jenner called it in his *Inquiry*. And these days, in the age of modern dairy hygiene, it is the house cat, "domesticated and caressed," that most often infects its fond human owners with this historic virus.

21

BENEFITS, RISKS, AND FEARS

"Antivaccine movements conform to a social dynamic that has to do partly with changes in society, partly with the internal dynamics of the neo-Luddite movement, partly with the very success of vaccines in eliminating or reducing feared diseases. Fear of science, dread of change, mistrust of anonymous corporate and government entities, and a nostalgia for a simpler real or imagined past all fed these movements in the past and help explain their resilience."

Arthur Allen, *Vaccine: The Controversial History of Medicine's Greatest Lifesaver*, 2007

In 1974 a research paper published under the aegis of the respected *British Medical Journal* described 36 children with epilepsy and linked their condition with routine vaccination against whooping cough. A number of British experts then expressed concerns about the proportion of adverse reactions to whooping cough vaccine, which included convulsions and a shocklike state. The vaccine in use at the time quite often caused fever and distress from which most infants recovered quickly, but concerns about the risks of permanent brain damage increased, and over the next few years vaccination rates in Britain fell sharply, reaching a low of 31 percent in 1978. Inevitably, whooping cough rates went up, together with the complications associated with this serious infection. In the early decades of the twentieth century, before the advent of a protective vaccine, whooping cough infected millions, and many thousands of children died. But with the disease controlled by immunization, the discussion about vaccine safety loomed large for every parent. When, in 1979, the time came for my own daughter to be vaccinated at just a few months old, I was no longer the dedicated young immunologist with a professional zeal for immunization. Instead I was an anxious father, asking questions and anguishing over the doubts surrounding this particular vaccine.

Uncertain what to do, I sought the advice of a colleague just appointed to the new post of Consultant Clinical Immunologist at Addenbrooke's Hospital, the Cambridge research hospital where I worked. His response was so measured as to be no help at all. With some trepidation we had our daughter vaccinated and she suffered no untoward effects. But concerns about this inoculation in combination with diphtheria and tetanus vaccines would soon be fueling antivaccine fears in America. And there *was* something mysterious about this vaccine.

Throughout the history of vaccines, episodes of harm caused by inoculation have jeopardized and sometimes forfeited the public's faith in this enormously successful practice. In the first hundred years of its use, Jenner's smallpox vaccine suffered disastrous episodes of contamination—with smallpox, with syphilis, with unknown agents causing jaundice. The problem continued into the twentieth century, when inadequately glycerinated smallpox vaccine in the United States became contaminated with *Clostridium tetani*. This germ, the tetanus bacterium, is killed by exposure to the air but had the chance to grow in covered vaccination wounds. This disaster in the early years of the new century spawned a resurgence in antivaccination sentiment in Philadelphia, leading eventually to the formation of a national antivaccination league. Thereafter, the U.S. Hygiene Laboratory, though never able to verify cases of tetanus contamination, sought to implement inspections of vaccine manufacturers and close facilities that didn't meet newly established standards.

But even uncontaminated smallpox is at the high end of modern vaccine risk. In America in the early decades of the twentieth century, compulsory vaccination against smallpox was militantly resisted in some states. Many citizens, like the townsfolk of Leicester in England a few decades earlier, had a natural antipathy to coercion. While threats to health abounded (diphtheria, whooping cough, measles, tuberculosis, infant diarrhea) the prevalent form of smallpox (*Variola minor*) produced only a mild illness. In the New York smallpox (*Variola major*) epidemic of 1947, when 5 million New Yorkers were heroically vaccinated, the inoculation campaign probably caused six deaths due to disseminating vaccinia in immuno-compromised people and a hundred or so serious adverse reactions in hypersensitive recipients. Since then, work on further attenuation of vaccinia virus has produced safer strains that grow only slowly in humans. Other poxviruses have been genetically engineered to persist when injected but not grow at all. But these experimental viruses have not been tested in humans. In the brief U.S. campaign to vaccinate police and health care workers against the threat of bioterrorism in the early twenty-first century, traditional smallpox vaccine was used.

In 1937, Max Theiler and his colleagues at the Rockefeller Institute developed a highly effective vaccine for yellow fever by passaging the virus in non-neural chick embryo tissue, selectively favoring strains that wouldn't grow in nerves. The viral cultures were maintained with a human serum supplement. This work earned Theiler the Nobel Prize in 1951. But in the first years of use, a tragic problem emerged: a small minority of recipients became ill with jaundice, signaling hepatitis. In Brazil, where a large vaccination campaign was running, many hundreds were affected, and nearly 30 died by 1941. Later it became clear where the problem lay: the vaccine was susceptible to contamination with an undefined infectious agent naturally present in the serum of some blood donors, and this mysterious agent was causing liver damage. Brazilians eventually solved the problem by substituting egg white proteins for human serum. Tragically, the problem was repeated in the United States. Scientists there carefully selected pooled human sera and heat-treated them to avoid the unknown jaundice-causing contaminant. Despite these precautions, the hepatitis agent crept through, and vaccinated U.S. troops were afflicted in 1941 in the early stages of World War II. As many as 10 percent of men in affected units were rendered unfit for duty in crucial theaters of combat. A serum-free vaccine became available in 1942. We now know that the agent causing serum-associated jaundice is a virus—hepatitis B virus.

Horse serum containing antibodies against diphtheria toxin, used to treat children infected with diphtheria, also suffered an episode of contamination early in the twentieth century, with deadly consequences. Horse serum can provoke an adverse reaction called serum sickness when the human immune system senses the horse protein as foreign and makes a defensive antibody. Molecular complexes that contain human antibody bound to the horse proteins sometimes form, and these cause rashes, fever, and malaise. In severe cases this is treated with corticosteroids. But a rare and much more serious problem proved devastating in October 1901, when a former milk-wagon horse, affectionately named Jim by his handlers, showed sign of tetanus infection and had to be put down. During his career Jim had provided several gallons of anti-diphtheria serum to protect children, but serum harvested in the short, silent period of tetanus incubation continued to be used, and this resulted in the death of twelve children in St Louis, Missouri. It was this tragedy, together with smallpox vaccine contaminated with tetanus, that led to the 1902 U.S. Biologics Act, the first in a long series of escalating standards introduced throughout the twentieth century. Despite the official records of these events, and the improvements that resulted, the personal stories of these children and their families remain largely untold.

The 1947 vaccine crusade against the smallpox outbreak in New York City harnessed the spirit of the war effort. It was this community-together ethos that made the campaign possible, and recently discharged air raid wardens and other wartime volunteers played crucial roles. These historic events, together with the example set by well-vaccinated troops, bolstered postwar public enthusiasm and belief in the power of vaccination across the nation, ensuring the continued success of the March of Dimes campaign against polio and culminating in the Salk vaccine trial in 1955. Despite the Cutter incident, the greatest modern disaster in mass vaccination (see chapter 14), the impetus of the crusade against polio ensured that a year later polio vaccination was well on the way to recovery. What no one knew was that throughout this period another insidious contamination was taking place unseen.

It was the great vaccinologist Maurice Hilleman (at Merck's facility in Pennsylvania, he developed vaccines for measles, mumps, hepatitis A and B, chickenpox, *Haemophilus influenzae,* and more) who first became suspicious of using monkey kidney cells to produce vaccines. He believed these organs potentially harbored unknown monkey viruses that might become passengers in vaccines. He and government scientist Bernice Eddy were the first to find such a virus in killed polio vaccine. This virus had been named SV40—the fortieth simian (monkey) virus to be identified. Hilleman also found the virus in live polio vaccine and raised the alarm at a conference in 1960. The problem was potentially serious because SV40 had been shown to produce tumors in laboratory animals. Albert Sabin firmly denied that any harm was possible, but Hilary Koprowski favored a switch to human cells for preparation of live polio vaccine. From 1962, manufacturers were required to certify their polio vaccines SV40 free, but between 1955 and 1962 a third of the nearly 100 million American children, and many millions of Soviet citizens, received polio vaccine that likely contained the monkey virus. The potential for harm on a prodigious scale alarmed some experts, but absolutely nothing could be done about it.

In 2002 the Institute of Medicine, the independent health arm of the U.S. National Academy of Sciences, reviewed long-term outcomes of polio vaccination and concluded that vaccinated populations did not show any increase in the risk of cancer. The evidence linking SV40 and tumor incidence was insufficient to conclude that the virus contaminating polio vaccine had caused cancer in any recipients. But, from the data available, it was also not possible to rule this out entirely.

Contaminating pathogens and live virus in killed vaccines are not the only problems for vaccines. A killed measles vaccine licensed in the United States in 1963 was responsible for a disorder that came to be called atypical measles syndrome. The defective vaccine was not withdrawn until 1967, and during that time more than a tenth of recipients, far from being protected, suffered measles infections more severe than normal. It eventually emerged that the chemical inactivation used did not preserve but rather destroyed an important viral protein whose function was to gain entry to human cells. Because no antibody was made to this component, vaccinated people were left more susceptible than normal, and when they encountered wild-type measles their disease was more severe. A live attenuated measles vaccine, developed by Maurice Hilleman, was introduced in 1968 and overcame the problem. Hilleman described the process of selecting weakened viruses in tissue culture as *rational empiricism,* his term for carefully judging the chance, undefined genetic changes that produce a weakened pathogen. Informally he called this trial and error method *guts and guesses.*

While problems with early killed measles vaccines were common, the Institute of Medicine has documented causal relationships between vaccines and much rarer medical conditions. In 1994 these included a rare allergic reaction to hepatitis B virus vaccine, thrombocytopenia (a blood clotting disorder), in 1 in 50,000 children receiving modern measles vaccination, and the ten or so cases of paralysis associated with live polio vaccination occurring each year in the United States. In 1995 the profile of this last problem was raised by parents of vaccine-damaged children, and the authorities instituted a switch from the live Sabin vaccine to a combination of killed Salk vaccine followed by live Sabin vaccine to prevent this annual toll. In 1998 they went further: Salk vaccine alone became the recommended regimen. The historic contest between Jonas Salk's famous vaccine and the globally successful live vaccine of Albert Sabin had at last come full circle in America.

Because all immunizations carry a measure of risk, albeit small risks in twenty-first-century vaccines, it is not desirable to continue vaccination once the chances of infection fall below a certain level. This was true of smallpox after its eradication from the developed world. At a meeting in Philadelphia in May 1965, the pioneering pediatrician C. Henry Kempe initiated what was to become a long debate on the value of continued smallpox vaccination in a land long free of the disease. He estimated that in the years since elimination in the United States (in 1949), several hundred children had died, and several thousand had been hospitalized, as a result of complications arising from smallpox vaccination. Safer (more attenuated) vaccines were being studied by

Kempe and others but were never licensed. The establishment, however, was concerned about sending the wrong message about vaccine safety to the American people and out across the globe. Kempe's concern went unheeded, though he continued his campaign.

Gradually, opinions changed, and in 1972 U.S. authorities revised their views on the risk of vaccination versus infection. Most states ended routine smallpox immunization. A more suitable policy of surveillance and containment with targeted vaccination, already used in Britain to counter imported outbreaks, was adopted.

The risk-to-benefit equation for vaccines has frequently been difficult for policy makers, and this was painfully true for U.S. authorities in the late 1970s, when fears of a new flu pandemic precipitated a national response. In February 1976 a New Jersey epidemiologist detected an unusual strain of flu in a soldier who'd died of respiratory illness at Fort Dix, New Jersey. The strain appeared similar to the deadly Spanish flu of 1918–19. Scientists at the CDC signaled their concern to the Department of Health and Human Services; in the run-up to a presidential election, this triggered a panicky government response. President Ford immediately asked Congress for funds to immunize the entire population against the threat of pandemic flu. In August, amid delays and difficulties with the manufacturing campaign, signs that the late winter outbreak was a false alarm were ignored in the rush to produce vaccine. The first outbreak of mysterious Legionnaires' disease in late July only fueled paranoia, and Congress rushed to assume liability for the vaccination program. By year's end 48 million doses had been given, but not a single additional case of this flu had been reported across America.

In November the first reports of an outbreak of Guillain–Barré syndrome, a neurological illness potentially linked to allergic responses to egg proteins in flu vaccine, began to appear, and eventually the government paid out millions in response to claims of vaccine damage. The harm apparently resulting from an unnecessary national immunization campaign did much to damage public faith in vaccination and discouraged investment in vaccines by the pharmaceutical industry. In 2003 the Institute of Medicine determined that there had been a small increased risk of Guillain–Barré syndrome following vaccination with the influenza vaccine in 1976. This increased risk was approximately one additional case per 100,000 people receiving the swine flu vaccine. Scientists have several theories as to why this might be, but the actual reason for the association remains unknown. The effect has not been seen in subsequent antiflu campaigns.

The debate around unnecessary and unwanted vaccination continued in the late twentieth century and is still with us today. In 1986 the world's first vaccine made by genetic engineering appeared. Maurice

Hilleman, university researcher William Rutter, and scientists at Chiron, a biotech company in San Francisco, produced a recombinant vaccine against hepatitis B (the virus that sometimes contaminated classical vaccines). DNA that encodes a key viral protein, readily recognized by the immune system, was inserted into yeast to make a pure vaccine product. This vaccine has a good safety record, but some parents remain uncertain about the desirability of compulsory use because the disease it prevents is restricted, in the Western world, to stigmatized groups such as drug abusers, prostitutes, and men who have sex with men. In the United States, vaccination is compulsory in most states. An alternative approach is to vaccinate at-risk groups against hepatitis B—the strategy used, for example, in the United Kingdom. But in public health terms, universal use of this vaccine has been very successful in reducing hepatitis, liver cirrhosis, and liver cancer.

Parental worries about vaccines understandably arise whenever an illness is perceived as minor and the vaccine against it is associated with adverse events. This was the case with the first vaccine against rotavirus, a universal gut pathogen first identified in 1973 that causes infant diarrhea. The illness results in several hundred thousand deaths worldwide in children under five but is less serious and readily managed in developed countries. The vaccine, ingeniously derived in 1998 from a similar virus that infects rhesus monkeys, was licensed for use in the United States and proved 80 to 100 percent effective at preventing diarrhea. But it soon emerged that the vaccine was associated with a rare bowel obstruction in one in every 12,000 vaccinees. The product was withdrawn and the experience provoked intense debate about the risks and benefits of the vaccine, inhibiting its use in countries where death rates from infection were high. It was not until 2006 that two new vaccines against rotavirus infection (produced by attenuating the human virus) were shown to be safe and effective in children. GlaxoSmithKline opted to establish its new vaccine in Brazil and Mexico where the need for it was high. In 2009 the World Health Organization recommended that rotavirus vaccination be included in all national immunization programs. The British government began to immunize all children in September 2013.

Another vaccine for a relatively mild disease, chickenpox, provoked debate when it was introduced in the United States in 1995. For the great majority of parents, the vaccine prevents a week or so of itchy discomfort for their small children. But at the public health level the vaccine prevents tens of thousands of hospitalizations for rare complications of the illness. Policy is different in the United Kingdom, where only people such as nonimmune health care workers are currently immunized.

Most modern decisions about when to vaccinate are finely balanced judgments of benefits, risks, and cost effectiveness, but now and then a storm gathers. On April 9, 1982, a TV program called *DPT: Vaccine Roulette* aired in America. The program graphically portrayed a story of serious damage to individuals and families caused by a version of the vaccine that had so concerned me as a parent in 1979. The vaccine in question, DTP, was a combination of three vaccines: diphtheria, tetanus, and pertussis. Pertussis is the name given to the whooping cough vaccine because whooping cough is caused by a bacterium called *Bordella pertussis*.

The pertussis component of the DTP vaccine had long been associated with adverse reactions. After the suspension of smallpox vaccination, pertussis became the most reactogenic vaccine given to children. Standards of safety in vaccination had steadily risen throughout the twentieth century, but in 1982 health authorities still did not actively seek out information on vaccine damage and no general program of compensation existed. As a direct result of the broadcast, parents who had long suspected that their children had been damaged by DTP came together, forming a pressure group called Dissatisfied Parents Together (DPT was the vaccine acronym used by the TV program producers).

The program's dramatic depiction of the lives of children whose severe disabilities it blamed on DTP led to a large increase in the number of lawsuits against vaccine manufacturers, and by 1985 companies had difficulty obtaining liability insurance. As a result, the price of the DTP vaccine increased dramatically, leading to shortages. The crisis eventually led to the passing of the U.S. National Childhood Vaccine Injury Act, approved by Congress in 1986, which established a federal no-fault system to compensate victims of injury caused by vaccines mandated by law. The majority of claims filed through this system, which gives the claimant the benefit of any doubt, have been related to injuries allegedly caused by DTP.

DTP has been one of the most problematic vaccines but it has now been replaced in developed countries by the safer version, DTaP, and this improvement exemplifies the quandary facing vaccinologists throughout the history of vaccination. It is the pertussis (whooping cough) component of the DTP vaccine that causes problems, provoking unpredictable, sometimes severe inflammation and fever in infants; the reason for this is that it contains whole killed bacterial cells. These cells are coated with a potent danger signal for the immune system, molecules called *lipopolysaccharides*. These hybrid molecules of carbohydrate and fat are widespread in bacteria but are not present in cells of the human body. They therefore provide a crucial warning signal to the primitive innate immune system by flagging the presence on an invader. Because their appearance can only mean the arrival of a pathogen, these molecules are powerful triggers of first-line immune defense, in which cascades of chemical messengers rapidly produce inflammation and fever. In DTP, the whole-cell pertussis element, as well as provoking a protective response to whooping cough, also acts as a nonspecific enhancer, or adjuvant, for the immune responses to tetanus and diphtheria.

The use of such components in vaccines is essential. Without natural or artificial adjuvants to jolt the immune system into action, the body fails to make a response strong enough for long-term protection. The quandary for vaccinologists is how best to trick the immune system into treating an innocuous vaccine as if it were a dangerous infection, and how to do it without provoking serious symptoms of disease. It is a balancing act. But all these adjuvant elements, alarm signals, and danger molecules present on pathogens were exploited in vaccines long before we understood what they were or how they worked. Vaccines cause harm when the reactogenic elements, useful up to a certain point, become dangerous beyond that level. Failures in the standardization and control of dosages in pertussis vaccines in the 1970s may have been responsible for vaccine damage before the safer vaccine was introduced.

By 1981 we had gained enough understanding for an attempt to improve the DTP vaccine. The strategy for improvement had its roots in work done by Yuji Sato and his colleagues in Japan. In 1976, they began identifying which individual components of pertussis whole-cell vaccine were protective against infection in laboratory mice. They then made a vaccine containing these subunits but excluded the highly reactogenic lipopolysaccharide components of whole bacteria. The problem with including lipopolysaccharide in a vaccine is that it's hard to control the degree of inflammation this potent danger signal provokes. A limited inflammatory response may be ideal for eliciting a strong immune response, but a higher level can be dangerous, causing fever, malaise, drops in blood pressure, and damage to tissues. As well as excluding this potentially dangerous molecule, Sato and his coworkers also worked out the right conditions for treating pertussis toxin (a protein essential for inducing immunity) with formaldehyde. The treatment inactivated the toxin but retained its shape to stimulate the immune system. Sato and his team had a strong motivation for their efforts—worries over serious adverse reactions to pertussis vaccine, including some fatalities, meant that in 1976 parental acceptance of the vaccine in Japan fell to little more than 10 percent. The great majority of infants went unprotected, and the number of whooping cough cases and associated deaths rose dramatically. The introduction of the cell-free, or acellular, vaccine, DTaP, in 1981 restored the public's faith, and DTaP was eventually licensed across the globe. DTP, which is cheaper than DTaP, is still used in some countries of the developing world.

In 1998 a vaccine paper was published in the U.K. medical journal the *Lancet,* and another storm began to brew. The British surgeon Andrew Wakefield and his co-researchers studied twelve children, nine of whom were autistic and suffered gastrointestinal problems and immunological deficiencies. They suggested that the onset of behavioral and gastrointestinal problems in these children was associated in time with the combination vaccine for measles mumps and rubella, commonly known as MMR. The authors speculated that the vaccine was a possible environmental trigger.

The study galvanized vaccine safety pressure groups and revived antivaccine sentiment in the United States. In England and Ireland the uptake of MMR vaccination declined, and the incidence of measles and its complications increased. In 2004 the *Lancet* retracted the report after Wakefield was found guilty by the U.K. General Medical Council of dishonesty and flouting ethics protocols in the conduct of the study. Although the U.S. Centers for Disease Control and Prevention declared the United States clear of measles in 2000, they later blamed reduced MMR vaccination rates for an outbreak of measles in 2008, in which 131 cases were reported, 11 percent of which were hospitalized.

In 2001 an expert committee convened by the Institute of Medicine reviewed the published data and concluded that the evidence favored rejection of a causal relationship between MMR vaccine and autistic spectrum disorders. They found that a consistent body of epidemiological evidence showed no association at a population level between MMR and Autism Spectrum Disorder. But they could not rule out the possibility that MMR vaccine could contribute to autism in a small number of children because existing epidemiological tools may not have enough precision to detect the occurrence of rare events. Other leading medical groups, including the World Health Organization and British health authorities, likewise concluded that there is no evidence linking MMR and autism. In a 2004 follow-up study, the Institute of Medicine repeated its categorical rejection of Wakefield's hypothesis. However, concerning an issue highlighted in Wakefield's study that is unrelated to MMR, they concluded that if some developmentally disabled children have previously unrecognized intestinal problems, treatment of these could lead to a better quality of life.

Another bête noire of vaccine safety pressure groups is the use of the mercury-containing antibacterial compound thiomersal as a preservative in vaccines (it is not present in MMR). As early as 1991, Maurice Hilleman at Merck raised the concern internally, pointing out that adding up the vaccines recommended for infants in the first six months of life resulted in a total mercury exposure greater than U.S. guidelines on consumption of mercury in whale meat, though the latter exists in a different chemical form. But it was not until 1999 that officials concerned with public safety initiated an urgent debate. While opinions were divided and the subject hotly debated, manufacturers were urged to remove thiomersal as a precautionary measure as soon as possible. Thiomersal (Merthiolate is a trade name for this compound) had been removed from all vaccines but one (influenza) by 2003. Subsequent studies have found no link between vaccine thiomersal and autism or any other neurological disorder.

Inevitably, these precautionary measures were seized upon by parents who blamed vaccination for their children's autism. Not surprisingly, the thiomersal story was damaging to public opinion and boosted antivaccine sentiment. Also not surprisingly, the cold currency of statistically validated scientific data counts for little among embittered and defiant parents for whom modern medicine can as yet do little.

While it is often impossible to convince some antivaccinationists with straightforward facts (American pressure groups continued to blame thiomersal for vaccine damage to children long after it was removed from childhood vaccines), the movement should be credited with driving important improvements. Campaigning groups critical of vaccines have driven up standards of surveillance and reporting for vaccine damage, the framing of compensation laws, and the separation between safety monitors and agencies implementing vaccine policy. More recently they won representation on the U.S. FDA vaccine advisory panel and helped drive reform in state vaccination exemption laws.

All those episodes of distress, anguish, injury, and death that have genuinely been associated with vaccination come down to the half dozen or so ways that vaccines can cause harm. Killed vaccines may be contaminated with live vaccinal virus; vaccines may carry hidden passenger viruses; components of vaccines may provoke adverse reactions, some of which are serious; live attenuated vaccines may, in a tiny minority of recipients, cause the disease they are intended to prevent; vaccines may simply be ineffective or they may be judged as unnecessary and unwanted by those compelled to receive them or have their children receive them; in immuno-compromised individuals, otherwise safe vaccines may overwhelm the weakened immune system.

Across the world there are many different national approaches to protecting populations by vaccination. In most U.S. states, childhood vaccinations are mandated by law. In Britain there is no compulsory vaccination; instead, an efficient state-run childhood vaccination system, which almost everyone subscribes to, ensures effective coverage. On a recent journey to one of Australia's remoter reaches, I asked my aboriginal guide if his community was well protected by vaccination against childhood diseases. "Oh yes," he told me. "We have a good clinic down in the valley." For a while he gazed at the river flowing through his ancestral land and then added wryly, "And if you don't take your kids down there they soon come after you."

22

INSPIRATION IN THE GLOBAL VILLAGE

"How can you put so much devilry
Into that translucent phantom shred
Of a frail corpus?"

D. H. Laurence, *the Mosquito*, 1923

J ust around the time when the last outbreaks of smallpox were being contained in the Horn of Africa, microminiaturization advances in the field of semiconductors reached the point where personal computers could be produced at an affordable price. The Information Age was dawning. Today we enjoy the full benefits of the digital revolution with instant access to knowledge for ordinary people in a way no one dreamed possible. Entire global financial systems are now impacted by the economy based on the manipulation of information, and we live in an information society made possible by near universal access to the internet.

Surprisingly, perhaps, the internet has its origins in the Cold War. In 1957 the Soviet Union launched the first man-made satellite, and in response U.S. President Dwight D. Eisenhower created the Advanced Research Project Agency (ARPA) and tasked it to produce countermeasures against a Soviet missile attack from space. With an urgent focus on communication, an Information Processing Techniques Office was set up, and this championed the benefits of a countrywide communications network. With the very highest national priority, lavish funding and generous resources were given to multiple university and industrial research organizations. In the rush to make the nation safe, the hallowed conventions of research were pushed aside. Even that great guardian of science, peer review, which every major publication since Jenner's *Inquiry* has had to undergo, was abandoned—ARPA could use discoveries *before* that lengthy process was complete. A special computer called an Interface Message Processor was developed, which made the communication network possible. ARPAnet went live in early October 1969. In 1990, the ARPAnet was transferred to National Science Foundation (NSF) control and soon became connected to the CSnet, which linked universities around North America, and then to the EUnet, which connected research facilities in Europe. In December 1990 a computer scientist at CERN, Tim Berners-Lee, implemented the first successful communication links enabling the creation of websites, making possible the development of the World Wide Web. Thanks to the NSF's enlightened management, and fueled by the huge popularity of the web, the use of the internet exploded after 1990. In response, the U.S. government transferred management to independent organizations to provide the internet we all enjoy today.

The transforming consequences of free and immediate global communication were foreseen by one man, the Canadian philosopher Herbert Marshall McLuhan. In the early 1960s, McLuhan wrote of the fundamental importance of media in shaping human sensibilities, predicting that the visual, individualistic print culture would be superseded by an age in which electronic media would replace visual culture with an immediate oral culture resembling primal tribal societies. He called this new society and its unifying identity "the global village."

One of the important consequences of the global village is that the conditions in which humans live across the planet can now be known with intimate immediacy in almost every living room and bedroom in the developed world. Human beings afflicted by drought, floods, famine, and disease are increasingly recognized as members of the global tribe compellingly portrayed in universal media. The power of the media, first seen in the March of Dimes appeals in 1950s America, has now been turned upon the wider world. Life in Asia and in Africa is on our screens, and in our thoughts, in a way it never was before; and if the children in our global village are dying of malaria, it's harder to ignore it. It's like a house on fire in your community. You can't stand by and let it burn; you have to pitch in and help fight the blaze. Standing right beside you is the biggest community of villagers in the history of civilization.

One important force in the worldwide campaign to deliver vaccines to those most in need is the Bill & Melinda Gates Foundation, whose focus includes preventable diseases, supporting new ways to fight and prevent the infections for which vaccines are already available, and funding work on vaccines yet to be developed. In 2000, at a time when the distribution of vaccines to children in the poorest nations was faltering, the Gates Foundation made a multimillion dollar commitment to GAVI (Global Alliance for Vaccines and Immunization), a public-private global health partnership that works to increase vaccination in the developing world through partnerships with governments, WHO, the United Nations Children's Fund (UNICEF), and the World Bank.

This new collaborative initiative has revitalized the Expanded Program on Immunization (EPI) established by the WHO in 1974 to build on the success of the smallpox eradication campaign. The vision of the twenty-first century campaign is to make it possible for all children in all countries to benefit from life-saving vaccines against measles, polio, whooping cough, tetanus, diphtheria, and tuberculosis. Recently, new and underutilized vaccines such as those for hepatitis B, bacterial flu, streptococcal pneumonia, rotavirus, and rubella have been included. Since Edward Jenner's discovery of smallpox vaccine in 1798, an ever-widening stream of human ingenuity has brought us new vaccines, new vaccine technologies, and a deepening understanding of the manifold cellular and molecular mechanisms through which vaccination works. We now have vaccines against more than 25 infectious diseases and a highly effective vaccine against one type of cancer (caused by a virus). But one of the greatest challenges of all is providing them to those most in need.

A major world study published in the *Lancet* in June 2010 showed that of the approximately 9 million deaths in children less than five years old in 2008, around 6 million were caused by infectious disease. Pneumonia was the biggest killer (18 percent), followed by diarrhea (15 percent), and then malaria (8 percent). The study investigated 193 countries, and half of the deaths occurred in five of them: India, Nigeria, Democratic Republic of the Congo, Pakistan, and China. Many could have been prevented by vaccination. The Fourth Millennium Development Goal of the WHO, UNICEF, Bill & Melinda Gates Foundation, and their partners is to reduce by two thirds, between 1990 and 2015, the under-five mortality rate.

It seems like a miracle—one of the hectic dreams of alchemists can truly be achieved. Base metals will never be transformed to gold; the elixir of youth will never be savored; but an end to plague and pestilence can sometimes come down to the specifics of a specialty in modern biomedical science. Vaccines have been successful against some of the most terrible diseases in human history, but others remain a challenge for the future. Foremost among them are malaria and HIV/AIDS. Why, in this age of powerful science, is it proving hard to develop an HIV vaccine? In most infectious diseases a good proportion of people make successful immune responses that defeat the pathogen naturally, but this doesn't happen with HIV. The immune system makes a wide range of responses, but none eliminates the virus and none stands out as being particularly effective. This make it hard to know what kind of immunity to aim for with a vaccine. Another obstacle is the lack of an accurate laboratory animal model of HIV infection to predict what might happen with a vaccine in humans.

The fact that HIV lives in immune cells adds a further complication. Some immune responses triggered by HIV actually help the infection. Certain interfering antibodies block neutralizing antibodies, while virus coated with so-called enhancing antibodies can enter immune cells more easily. HIV hoodwinks the immune system by reducing MHC molecules on cells it infects, which would otherwise flag them up to killer immune cells. Another problem is the diverse nature of HIV viral subtypes: more than a dozen exist with different geographic distributions. Constructing a vaccine to protect against them all is an enormous challenge. Yet another obstacle is the subtle changes in the virus within individuals during the chronic infection, which keeps the virus one step ahead of the immune system. HIV can also hide in the DNA of the infected host as a so-called provirus. Until we have a vaccine, the only weapons against HIV/AIDS are public health measures, education, and complex regimens of antiretroviral drugs.

In Africa, the HIV pandemic spread gradually across a much older, but equally deadly, established pandemic: malaria. This disease was named for what was thought to be its cause: *mal'aria,* the foul air exhaled by the swamplands of Italy. British army surgeon Ronald Ross, working in Secunderabad, India, in 1897, was the first to show that malaria is transmitted by mosquitoes. He was able to see malaria parasites under the microscope in a mosquito he'd deliberately fed on a malaria patient. Malaria is one of the major causes of mortality among infants and children in sub-Saharan Africa. So how can we extend the miracle of vaccination to malaria, HIV/AIDS, and other diseases against which we seem to have little or no defense?

Vaccination is an extraordinary trick played on the immune system, and Jenner's archetypal vaccine illustrates this well. A relatively innocuous living infection, that by chance resembles a virulent one, provokes long-lived protective antibodies against both diseases. But the relationship that Jenner found is rare. Another successful strategy is to use living but attenuated infectious agents in vaccines. These also fool the immune system into producing a long-term protective response against the full-blown disease. The same is true for killed vaccines, but here, to make the trick work, we have to add adjuvants that make the harmless vaccine look dangerous. The adjuvants in traditional vaccines were discovered by chance, but at last we are beginning to understand how they work, and we can artificially reproduce the danger signals that accompany deadly infections.

It is the evolutionarily primitive innate immune system that recognizes danger signals, and these are usually general molecular motifs characteristic of germs but not present in human cells (or those of animals). Synthetic versions of these danger molecules can be added to purified key proteins of viruses or bacteria to ensure the immune system is tricked into mounting a full response, as if the appearance of the pure protein were a genuine infection. Alternatively, the intermediate molecular messengers sent by the innate immune system to alert the adaptive immune system to danger can be added to vaccines.

When Hilary Koprowski and Albert Sabin attenuated their polioviruses, they did so by trial and error, repeating the weakening effect of transfer in tissue culture until the virus was rendered safe. But we now know a great deal about the specific genes that make a virus virulent, and these can be deliberately altered or deleted to tailor-make an attenuated vaccine. Some of these engineered viruses are designed not to grow in humans. These nonreplicating agents, which still possess their danger signals, can also be used to carry the genes of unrelated infectious agents to make new vaccines. One such modified virus is none other than vaccinia, the classic virus used to eradicate smallpox. New versions, safer than the original, have now been engineered and can be loaded with the genes of other pathogens to produce live vaccines for the future.

But in the case of malaria and HIV/AIDS, we still need something more. Successful vaccines are much more readily achievable against pathogens that express long-lived antigens and that are neutralized by antibodies. Smallpox and polio make good examples. But pathogens that change their antigens, and that succumb only to a combination of both cellular and antibody immune responses, are much more difficult to tackle. HIV and malaria fall into this category.

For HIV, early attempts to create a vaccine failed in the 1990s and early in the new millennium because of the problem of viral diversity and limitations in the breadth and strength of responses. The failures included a new vaccine aimed to induce cell-mediated immunity to conserved antigens—elements that do not vary so much between strains or during chronic infection. Cell-mediated immune responses, which include direct killing of virally infected cells by lymphocytes, are likely to be an essential component of any effective immune response against HIV. (So too are immune responses protecting mucus membranes of the intestine and reproductive tract.) Exploiting the latest principles, this HIV vaccine consisted of a nonreplicating viral carrier for HIV proteins. This carrier is called *adenovirus,* and the widespread natural version of this virus causes mild respiratory infections in humans; but the nonreplicating version used to make the vaccine does not cause disease. This engineered virus carried genes encoding key HIV proteins essential to the function of HIV, and the experimental vaccine had shown itself to be effective in nonhuman primates. Sadly, it did not work in the 3,000 human volunteers who received it. But in 2009 a vaccine trial exploiting an additional new principle gave hope for the future. This new vaccine consisted of a live poxvirus carrier (canarypox) carrying genes encoding key HIV proteins for the first immunization. In contrast, the second dose of vaccine was quite different: it contained not genes but purified (nonliving) HIV proteins mixed with a conventional adjuvant called alum. By giving the vaccine antigens in these two very different guises (sometimes called a heterologous prime-boost regimen), the intention was to access both cellular and antibody arms of the immune response. This vaccine, given to 16,000 volunteers in Thailand, yielded just over 30 percent protection against HIV infection. While this is a modest effect, some see it as encouraging for the future. Other trials, using different prime-boost regimens with various HIV elements and assorted alarm signals, are in the planning phase.

Along with HIV and tuberculosis, malaria is one of the big three killers in the developing world and is estimated to have killed more humans throughout history than any other single disease. Until recently, the only modern weapons against it were insecticide-impregnated bed nets and highly effective combination drug regimens to treat the infection. Malaria is caused by several species of a protozoan (single-celled) parasite called *Plasmodium,* and the parasite's complex life cycle, which involves both mosquitoes and humans, makes it very difficult to create a vaccine. When the infected female mosquito bites to get a blood meal, she injects wormlike parasites (called sporozoites) from her salivary gland into the skin. When the parasites arrive in the liver, they multiply many thousand fold without causing symptoms. A vaccine for this first stage of the life cycle would therefore block infection and prevent malarial sickness. Liver cells packed with parasites then burst, releasing showers of the next parasite stage (merozoites), and it is these circulating forms that invade red blood cells and cause sickness. The cycle begins again when sexual forms derived from this parasite stage (gametocytes) are taken up by a feeding mosquito, where they subsequently mate to generate more parasites whose numbers increase at every stage in their development.

In endemic areas of Africa (and in parts of Asia and Central and South America), malaria causes only a mild illness in adults because their acquired immunity limits the infection. Newborns are protected by maternal antibodies in breast milk and also because the structure of fetal red blood cells doesn't suit the parasite. The vulnerable population therefore consists of infants and children. Mothers in their first pregnancy are also susceptible to malaria.

Because of the complexity and diversity of the malaria parasite (so much bigger and more complicated than viruses or bacteria) and its multistage life cycle, vaccine discovery is extremely challenging. However, even a partially effective vaccine should diminish parasite transmission and reduce sickness and mortality. But in order to do this the vaccine will still have to outperform the immunity stimulated by natural infection. This is because acquired adult immunity protects against illness but does not prevent infection. Most vaccines, at best, merely reproduce the immunity caused by an episode of the disease, as is the case for polio and smallpox. But in malaria, this isn't good enough to stop transmission. What's needed is a supervaccine incorporating potent danger signals—a formidable task indeed. But such a vaccine is already on the horizon.

The history of malaria vaccination is rich in heroic experiments. A study in 1973 by collaborators in Maryland and New York went to remarkable lengths to show that vaccine-induced protection against malaria was possible. Three human volunteers were each exposed to the repeated bites of several hundred mosquitoes that had been irradiated to weaken the *Plasmodium* parasites they were carrying. The three men were then exposed to untreated mosquitoes bursting with vigorous parasites. One of them was completely protected by the immunity generated by the live attenuated parasites, and this set in train four decades of intrepid research, much of it funded by the U.S. Army, whose renowned Walter Reed Army Institute did a great deal of work. By 2009 this line of investigation had reached its zenith in the laboratory: certain parasite genes were selectively knocked out so that the parasites could do everything but grow in the human liver. Used as a live attenuated vaccine to give exposure but cause no sickness, these parasites should induce immunity to sporozoites and block subsequent infection. Such a vaccine candidate would need a great deal of further development to make a product for large-scale use: the first clinical trial will require volunteers to receive the vaccine from the bites of infected mosquitoes—a mixture of cutting edge science and basic expedience familiar in vaccinology. Other researchers have developed a standardized live sporozoite vaccine attenuated not by gene knockouts but by fine-tuned irradiation. But both these live attenuated vaccines must await the development of a shallow needle delivery system to mimic those devilish mosquito bites.

Others investigators have pioneered a quite different path by using purified proteins of malarial sporozoites synthesized artificially in yeast. The engineered proteins stimulate both cellular and antibody-mediated immunity and are ingeniously combined with a virus-like carrier particle from a hepatitis B vaccine (because it looks like a virus the immune system pays great attention to it). This vaccine stimulates the production of antibodies to sporozoites, which prevent the parasite from gaining entry to liver cells. Because the nonliving proteins do not appear dangerous to the immune system, danger signals (those generic molecular patterns characteristic of pathogens) are added to convince the immune system it is dealing with a dangerous invader. Many agencies have contributed to the ongoing collaborative studies—including U.S. Army researchers and the PATH Malaria Vaccine Initiative, a nonprofit organization supported by the Gates Foundation—but the vaccine was created and developed by GlaxoSmithKline's Biologicals Institute at Rixensart near Brussels.

On October 19, 2011, the first results from a large-scale trial were announced at the Global Malaria Forum in Seattle, Washington. They show a halving of malaria episodes during the first year after immunization—a source of some satisfaction for the project's originator, Dr. Joe Cohen. More than 15,000 infants have been vaccinated in seven sub-Saharan African countries—Burkina Faso, Gabon, Ghana, Kenya, Malawi, Mozambique, and Tanzania—by teams led by African scientists. If safety and efficacy continue to be satisfactory, this vaccine will be used against malaria in Africa—the first ever vaccine against a parasitic disease.

But another historic war is being fought against disease in Africa. And the same campaign is being waged in Asia. Since 1988, more than 2.5 billion children have been immunized against polio through the Global Polio Eradication Initiative (GPEI). This great endeavor has taken place in more than 200 countries, helped by 20 million volunteers. Back in 2006, only four countries remained that had never stopped polio transmission, and annual case numbers had decreased by over 99 percent. But since 2003, annual global cases have fluctuated frustratingly between 1,000 and 2,000, and between 12 and 23 countries every year have reported polio cases. In 2008, the WHO called for an intensified strategy to eradicate polio from the remaining affected countries and rid the world of polio for all time. At the end of 2009, the campaign instituted a three-year strategic plan: an aggressive, time-bound strategy with measurable milestones to banish polio. One of the three types of polio (type 2) has already been eliminated. The campaign is led by national governments in partnership with four agencies: the World Health Organization, Rotary International, U.S. Centers for Disease Control and Prevention, and UNICEF. Funding partners include the Bill & Melinda Gates Foundation, the World Bank, donor governments, the EU commission, and the International Red Cross and Red Crescent societies. Soon, perhaps very soon, polio is likely to become the second great infectious scourge of humankind to be defeated by vaccination.

23

A TEAM OF MANY COLORS

"It is the same moon that wanes today that will be the full moon tomorrow."

Nigerian proverb

I n January 2012 there was much to celebrate for the global campaign to eradicate polio. For the first time in history, a year passed in India with not a single child becoming paralyzed. In the whole of 2011, India had only one case, a two-year-old girl in Howrah, close to Kolkata, West Bengal. There were 42 cases in India the year before. A case with onset of paralysis on July 30 in western Kenya was the only case to be reported in the Horn of Africa in 2011. The UN Secretary-General Ban Ki-moon wrote to the heads of state of Nigeria, Pakistan, Afghanistan, Chad, and the Democratic Republic of the Congo, the regions still afflicted, about the urgent need to complete the eradication of polio.

Since 1988, the Global Polio Eradication Initiative (GPEI) has reduced the number of polio cases worldwide by more than 99 percent (from 350,000 new cases in 1988 to just 223 in 2012), but this dramatic reduction has not followed a simple downward path, particularly in the crucial approaches to the "final push." Because the poliovirus is spread largely through fecal contamination of food or water, it remains a particular problem in parts of the world where good sanitation and access to clean drinking water are limited. Also, the fact that 95 percent of infections are without symptoms (though these cases can still spread the infection) makes polio difficult to detect but all too easy to disseminate.

The Federal Republic of Nigeria, the most populous African country and the seventh-most-populous nation in the world, has long been seen as the key to defeating polio in Africa. This great country, at the heart of West Africa where the continent begins its westward sweep, remains one of the most stubborn reservoirs of poliovirus in the word. Polio has continued to be transmitted among people in seven of Nigeria's northern states; this creates a wider problem because the region has been the source of virus invading at least twelve African countries previously made polio-free, repeatedly dashing hopes of an end to the disease. The founding vision for global eradication of poliomyelitis predicted completion in 2000, and when this failed a new target of 2005 was instituted. Both dreams ended in Nigeria.

The campaign in Nigeria underscores the complexity of the global endeavor against polio and tragically illustrates the fragility of gains already made. The all but insurmountable problems confronting the campaign are woven through the fabric of Nigerian life. How is it, one might ask, that the campaign in Nigeria falters, even as other African nations with fewer resources succeed in ending transmission of the disease? One problem is Nigeria's size. With 155 million people in 36 states (some as big as other African nations), the challenges of confronting polio in the face of poverty, corruption, weak institutions, social conflict, and political turbulence are magnified. In addition, the effort is at a national level while Nigeria has traditions of regional, noncentralized government. Certain cultural particularities can also be a problem: what the West might see as widespread criminal corruption is frequently a part of prebendary customs in which elected officials believe they have a right to personal shares in the revenues of government.

Another obstacle to success is Nigeria's diversity. The nation encompasses more than 250 ethnic groups, more or less divided equally between Muslim and Christian faiths with a small minority observing traditional religions. The principal cultures of what became Nigeria in the late nineteenth century were united only by colonial decree since no natural commonality existed between them. Nigeria is a country of many landscapes: the dry, open grasslands of the savannah make cereal farming and herding a way of life for the Hausa and the Fulani peoples of the north. The Hausa language ranks among the world's great languages with a rich heritage of music, poetry, and prose. The Fulani people are the largest nomadic group in the world; herders and traders, their culture values bravery and beauty in its traditional customs and styles of dress and decoration. The lush tropical forests to the south yield fruits and vegetables, providing the livelihoods of the Yoruba and Igbo peoples. Along the coast, small ethnic groups such as the Ijaw and Kalabari build elevated houses in a land of creeks, lagoons, and marshes where fishing and the trade in salt provide a living. In the center of the country, the Niger and Benue rivers come together to create a great "Y," which forms the boundaries of the three major ethnic groups: the Hausa in the north, the Yoruba in the southwest, and the Igbo in the southeast.

The Hausa and Fulani cultures of the north are Muslim, while the southern half of Nigeria is largely Christian. When the campaign against polio was first launched in 1996, Nigeria was in deep economic and political crisis. After a string of armed coups, the brutal and corrupt regime of General Sani Abacha took control of the nation. As a result, corrosion of state institutions, suppression of civil society, and economic stagnation accelerated the long decline from the boom years of the 1970s, when oil made Nigeria rich. Autocratic revolutionary governments can sometimes benefit immunization campaigns, as in the case of smallpox in Ethiopia when the power of the new regime lent impetus to the vaccination effort in 1976. But in Nigeria, health care, which included immunization programs, was transferred to poorly resourced local authorities with all the problems of fragmentation and underfunding that this entailed.

The effect on delivery of public health was calamitous. Immunization coverage plummeted to around 30 percent and didn't much improve when General Abacha brought vaccination services back under national control. His National Program on Immunization (NPI) was largely ineffective, and General Abacha's wife, according to retrospective reports by independent monitors, was repeatedly implicated in corruption scandals involving vaccine procurement. The NPI director between 1996 and 2005 was finally accused of gross incompetence and corruption.

With the death of General Abacha in 1998, a precarious transition to civilian democracy allowed progress to be made once more. Under the leadership of Olusegun Obasanjo, a Christian former military leader from the south, the target of eradicating polio gained ground despite the chaotic legacy of the military era. Hopes were raised that the 2005 target for defeating polio might just be possible. Major progress against polio was achieved in the southern regions—the power base of the democratic government—and this sector of the country also enjoyed much greater economic recovery. Traditional enmities between north and south, always present in Nigerian national life, were soon reflected in the vaccination campaign. Northern disaffection with the more affluent south, and with the wider world whose resources seemed always to flow to the south, fostered distrust and suspicion toward interventions by Western nations.

All these problems came together in the polio vaccination campaign. As the eradication effort intensified, some northern communities began to ask questions about the polio vaccine. Why was it that polio vaccine was provided free of charge when medicines for far more common and more deadly illnesses (malaria, cholera, diarrheal diseases) were expensive and not readily accessible? And why were communities so earnestly entreated by Western agencies and government from the Christian south to receive this particular vaccine when everything else remained in very short supply? Not surprisingly, rumors sprang up, some fueled by local imams, that the immunization program was a Western plot to control or eliminate Muslim populations in the north by causing sterility or by spreading the HIV virus.

The atmosphere of distrust and suspicion was worsened by an episode fresh in the memories of Nigerians in the north. In the midst of a major meningitis outbreak in 1996, the American pharmaceutical giant Pfizer launched a clinical trial of the antibiotic trovaflaxin in Nigeria to see if it was better than existing treatments. Eleven children participating in the trial died, and Pfizer faced accusations of failing to obtain fully informed parental consent. The controversy sparked an outcry and provoked street demonstrations in Kano, the second-largest city in Nigeria, and spurred demands for greater accountability. The deaths and the subsequent storm over responsibility and compensation had an enduring impact on the already suspicious peoples of northern Nigeria. As the new millennium arrived, more trouble lay ahead for the campaign against polio.

In July 2003 opposition to the vaccination program in the north of the country intensified. The leadership of both the prominent Muslim umbrella group, Jama'atul Nasril Islam, and Nigeria's Supreme Council for Sharia questioned the polio vaccine's safety and encouraged a boycott. The incumbent president of the council went further, publicly claiming that there were reasons to believe the vaccine was contaminated with antifertility drugs, viruses that cause HIV/AIDS, and a simian (monkey) virus likely to cause cancer.

In response to the rising alarm, Ibrahim Shekarau, then governor of Kano, took the decision to suspend administration of the vaccine pending an investigation of its safety. It didn't take long for the neighboring states of Bauchi, Kaduna, and Zamfara to follow suit. Some observers at the time suspected that leaders saw political advantage in demonizing federal authority coming from the south. As news of the boycott spread, feelings ran high, and anxious parents began refusing entry when vaccinators called at their homes. Some marked their doors with fake official signs reserved for households already vaccinated. Others painted their children's fingernails to mimic the official marker vaccinated children receive.

The GPEI, the UN, and the U.S. government scrambled to resolve the escalating crisis with multilevel diplomatic initiatives, including high-level overtures to senior Nigerian political figures and religious leaders as well as to the Organization of the Islamic Conference. The most powerful sponsors of the campaign began a dialogue with African Union leadership urging African nations to put pressure on Nigerian state authorities to resolve the situation. At the same time, arrangements were made to source polio vaccine produced in the largely Muslim country of Indonesia to restore confidence in its purity. The scare stories were without substance, but they gained some credence with Nigerians who had heard stories of the early development of polio vaccine, when SV40 virus contaminated Salk and Sabin vaccines.

As the diplomatic pressures, applied through repeated representations at the international level, continued to build, the WHO made sure that Saudi Arabia was fully aware of the risks of polio crossing its borders during the *hajj*, the great yearly pilgrimage to Mecca. The Saudi government responded, and officials warned the authorities in Kano that pilgrims would be required to present vaccination certificates and also receive a further vaccination at the border before entering Saudi Arabia. Good sense and better understanding eventually prevailed, and the boycott in northern Nigeria was brought to an end in 2004.

Something about the way in which the trouble was resolved then had a magical effect: within a year the very same religious and civil leaders had more than made up for their former resistance. In a powerful public gesture, the Kano State governor, Ibrahim Shekarau, allowed President Obasanjo to publicly administer the vaccine drops to his one-year-old daughter. In 2006, the newly appointed Sultan of Sokoto, Muhammadu Sa'ad Abubakar III, the spiritual leader of Nigeria's 70 million Muslims, became a great advocate of polio immunization. The ancient city of Sokoto is rich with the history of the old Sokoto Caliphate that established Islam in this region of West Africa, and the Sultan's words carry great authority for all Nigerian muslims. Nigerians did not relish their reputation as the nation reinfecting Africa, and the crisis lit a beacon of renewed national determination that burns today. A spirited and determined effort to halt the spread of polio and fight for its extinction at the federal, state, and local community level got underway. Since 2008 this has seen greater government contributions to the campaign to raise awareness and implement the most effective strategies for eradication. When President Obasanjo stepped down in 2007, his successor, Umara Yar'Adua, a long-standing supporter of vaccination, took up the national cause; since his death in 2010, the new leader of Nigeria, President Goodluck Jonathon, has been a committed supporter, pledging an additional $60 million in federal funds to polio in October 2011.

Events following the vaccine boycott sadly illustrate the readiness of poliovirus to return whenever vaccination pressure weakens. Confirmed cases of polio paralysis rose from just 202 in 2002 to 1,122 in 2006, and the virus spread to neighboring countries and beyond. But by

2010 religious leaders in Nigeria were at last united in backing the vaccination campaign. As in all countries being targeted, the most important weapons are Supplementary Immunization Activities (SIAs). These are vast operations to deliver vaccination to every household irrespective of vaccination history in a "catch-up" operation. They often take place on National Immunizations Days, which aim to raise awareness. Amid great fanfare and razzmatazz, everyone is made aware of what is happening, which helps ensure that no one falls through the net. These heroic operations are mounted in addition to the national routine immunization of children in the first year of life. In 2011, Nigeria had only 62 cases of polio paralysis. Early in 2012, the much-loved folk singer Dan Maraya Jos, appointed as the "End Polio Now" Rotary Ambassador, toured the high-risk areas to drive home, through traditional song, the importance of immunization. One of his most famous songs (about threats to a happy marriage) condemns the spell-casting alternatives to modern medicine.

The global initiative to eradicate polio is a crucial part of the larger global campaign to carry vaccines to every corner of the globe. Where polio vaccination reaches, wider benefits can also be delivered in terms of other vaccines and access to health-care resources. Awareness in target communities is vital, but so too is the awareness of those in donor countries. In 2012 the much-admired Scottish actor Ewan McGregor traveled as a UNICEF ambassador to Nepal and the Republic of the Congo, with teams carrying vital vaccines to two of the remotest destinations on earth. Because some vaccines are damaged by heat and by freezing, these vaccination teams take care to transport them at temperatures between 4 and 8°C—the vaccine cold chain. Among them is the polio vaccine. From a storage center in Patna, Bihar State, northern India, where the team picked up supplies, the mission made its way to the first destination, a camp for nomads in Patna itself where many children were immunized. At Patna Railway Station, McGregor met Amita, a young worker and a victim of childhood polio who was dedicated to seeking out children on the platforms and giving them the oral vaccine. The expedition continued by motorbike and boat, reaching the village of Terasi in Bihar after a twelve-hour journey. Here a little girl named Phultan received her vaccination. The next stop was the border with Nepal, a crossing point for some of the 2 million people who travel between India and Nepal each day. Here, in the border town of Forbesganj, the passage of each day also sees several hundred children immunized.

The next phase of their journey ended when their small plane landed above Kathmandu on one of the world's most daunting runways, where wind and snow often prevent the delivery of vaccines. To ensure no infant misses out, UNICEF and its partners maintain a vaccine stockpile here sufficient for nine weeks. Without a pause to wonder at the shrines, palaces, and gilded stupas of this ancient and mysterious city, the expedition pressed on. McGregor set out on foot in the company of Shersingh, a Nepalese UNICEF health worker and intrepid veteran of this journey without roads. Two days later they arrived at Luma with their insulated box of vaccines. At this remote village on the mountainside, four-month-old Bura received her measles vaccination. Another trek took them to their final Himalayan destination, where eleven-month-old Nirmala received her measles shot. As UNICEF organizers point out, it had taken four days to get the vaccine to her; it was all over in a second, but vaccination should protect Nirmala for a lifetime.

McGregor also joined UNICEF workers on a cold-chain expedition that set out from Brazzaville in the Republic of the Congo. This African mission seemed a darker journey, bound for destinations even more remote. In the words of Joseph Conrad, McGregor's progress up the Congo River seemed like "travelling back to the earliest beginnings of the world, when vegetation rioted on the earth and the big trees were kings." And the mission continued, as it did for Conrad's hero, "with wonder amongst the overwhelming realities of this strange world of plants, and water, and silence." Some things McGregor encountered were troubling, to say the least. One tribe, indigenous autochthons, appeared to be enslaved by the local Bantu people through customs no one questioned. Yet these things were tangential to the vaccination story, and in the end it didn't matter: this was no journey to the heart of darkness, and characters who seemed oppressed retained their self-belief, proudly delivering their children from the depths of the forest to be vaccinated, a right questioned by no one. Beyond this, the journey upstream continued, and at last they reaching the Baka tribe, hunter-gatherers who dwell deep in trackless and uncharted forest. Here, local messengers carried the news of the vaccinators' visit to those otherwise impossible to reach. The Baka people then made the journey to the larger villages so their children could be vaccinated. There *are* peoples more remote than these: the tribes of the Amazonian forest and elsewhere as yet uncontacted by global civilization. But for now, at least, these human beings remain strangers to us and, hopefully, to our diseases.

Outside of Africa, polio remains a problem in both Afghanistan, which reported 80 cases in 2011, and Pakistan, which had 198. Because of the ready transmissibility of polio, officials have had to revise their strategy from the surveillance-containment method, which was so successful against smallpox. It seems population immunity of greater than 95 percent is required to halt transmission in Asia, while transmission in sub-Saharan Africa appears to cease when immunity exceeds 80–85 percent. In Asia, persistent transmission is highly localized in a few densely populated areas, so the approach is district-specific and intensive, with frequent supplementary immunization to boost population immunity above 95 percent. In Africa, wider multinational strategies are implemented. In both campaigns polio surveillance remains vital. The first sign of polio is a floppy paralysis with sudden onset; these cases are reported through a network of monitors, and stool samples are taken so that tests for the virus can be conducted in central laboratories.

The vaccine used in the eradication drive is the oral polio vaccine developed by Albert Sabin. The intensive, large-scale vaccination campaigns essential to defeating polio are only possible because this vaccine is so easily given and produces long-lasting immunity, often with a single dose. One consequence, though, is that in very rare cases the virus reverts to its original virulence, causing paralysis (estimates of risk over the period in which this vaccine has been in use have varied between 1 in 0.75 million and 1 in 2.7 million). Another drawback is that people with severely weakened immune systems may shed the vaccine virus and infect others. The vaccine virus can therefore circulate in areas with low rates of vaccine coverage. The WHO is developing a strategy to deal with this and will switch their limited resources to the problem once global eradication has at last been achieved. Those individuals who suffer vaccine-induced paralysis pay an enormous price for the success of the eradication campaign, but in the future new concepts in vaccine design may overcome this problem. Polio is a relatively simple virus with an RNA core and a protein coat. By using the high energy X-rays that are produced when electrons are accelerated in state-of-the-art research facilities, scientists can analyze the structure of viruses in very great detail. This provides precise information for making synthetic protein viral shells empty of genetic material. Used in a vaccine, these highly accurate virus models, which are noninfectious and cannot become virulent, trick the immune system into mounting an immune response. This new technology may provide an entirely safe polio vaccine for the future. The campaign against polio has been enormously successful. Since 1988, more than 2.5 billion children have been immunized thanks to the unprecedented cooperation of more than 200 countries and 20 million volunteers, backed by an international investment of more than U.S. $8 billion. Annual case numbers have decreased by over 99 percent. But problems can still emerge without warning to jeopardize any final victory, even in countries once free of polio.

In the spring of 2010, a symptom-free traveler arrived from India in Tajikistan, a small landlocked country of lakes and snow-capped mountains in Central Asia. The traveler was infected with polio. The disease quickly spread. By mid-May, 108 Tajik children had been infected and twelve of them had died. Mass vaccination campaigns were quickly mounted across the country and in neighboring Uzbekistan to try to contain the virus, the first epidemic in a decade. Within a month, thousands of health workers vaccinated at least 97 percent of the children in each region of this mountainous land as the Ministry of Health, assisted by UNICEF and the WHO, rolled out the emergency campaign. In May Russian authorities reported a case of polio in a child who had traveled from Tajikistan. The Tajik outbreak represented 75 percent of the world's polio cases, far exceeding numbers in India and Nigeria and three times the number in Pakistan. Eventually, 460 people were infected, and countries as far away as Canada became concerned by the fact that any Western country with low vaccination rates is only one asymptomatic air-traveler away from an outbreak. But the emergency campaign in Tajikistan quickly prevailed: there were no more cases after July 4. Since then the country has remained free of polio.

This chapter on the recent history of polio eradication highlights the seriousness and unpredictability of polio importation and the ease with which long-term achievements can be suddenly reversed. Importation continues to be a major threat to campaign success in West Africa, where nomadic traditions threaten to carry infection to Nigeria's neighbors. In these impoverished countries a large-scale, rapid response of the kind achieved in Tajikistan would be difficult to mount.

So when will global eradication finally happen? When will it be, "too manifest to admit of contradiction that the annihilation of this dreadful scourge," as Edward Jenner might have put it, will at last be achieved? Global recorded cases dropped as low as 483 in 2001 and then sprang back to over 1,000, remaining near that level for the next nine years. But in 2008 the Global Polio Eradication Initiative (GPEI) launched an intensified effort with renewed commitment by leaders and donors. Their strategic plan for 2010–2013 identified targets toward interrupting wild poliovirus transmission globally by the end of 2012, and although this target was not achieved, the fresh impetus has established a new vision for the endgame, a new sense of urgency, and a new focus on performance and the realization of targets. In total 223 new cases were recorded in 2012. The plan also specifies that three years of surveillance with zero cases will be necessary after transmission has been interrupted in order to certify an end to poliomyelitis.

In Geneva, Switzerland, on May 24, 2012, the GPEI announced a crucial change in its drive to put an end to polio. The whole endeavor shifted into higher gear with an Emergency Action Plan intended to boost vaccination coverage in Nigeria, Pakistan, and Afghanistan, the three remaining polio-endemic countries since India succeeded in eradicating endemic polio early in 2012. (With no new polio cases in the last two years, India is now classified as polio-free.) The target of the plan is to achieve immunization levels of 80–85 percent in Africa and 95 percent in Asia—levels estimated to be sufficient to stop polio transmission. A simultaneous resolution to declare polio a programmatic emergency for global public health was adopted by health ministers at the World Health Assembly in Geneva, elevating its priority against other programs. The initiative included new approaches tailored to each country to improve campaign performance, heightened accountability, coordination and oversight within and between collaborating agencies, and a surge in technical assistance and social mobilization to ensure support for vaccination. Polio eradication was judged to be at a tipping point between success and failure, and some estimated that failure at this juncture could lead to a widespread resurgence of polio within a decade. Success, on the other hand, could generate net benefits of tens of billions of dollars globally by 2035, with the bulk of savings in the poorest countries. The calculations were based on investments made since the beginning of the GPEI campaign and savings from reduced treatment costs and gains in productivity.

Around the time these changes were announced, events in Africa were tossing up new challenges. Mali in West Africa is a land of contrasts, from the lush tropical forests in the south to the vast aridity of the Sahara in the northern regions. The country is divided by the Niger River, which rises in the southwest Fouta Djallon highlands and flows 1,000 miles through the land, swelling when the rains arrive to stretches a mile wide. But the Niger is not the only division in this nation; the peoples of Mali are as diverse as its landscape. In January 2012, after a decade of instability, fighting in Mali led to a coup and takeover by jihadi groups from the Maghreb region to the north of Mali. Hundreds of thousands of people, fearing the harshness of the new regime, fled to neighboring lands. They trekked across the desert, some with donkey carts piled high with their household belongings. Among their worst fears was the possibility of Western intervention to dislodge the Islamic militants. They dreaded the bombings they'd heard about in Libya only the year before. For many, makeshift dwellings were the only shelter through the rainy season. Tens of thousands of refugees survived in *baches*—simple dwellings of sticks and canvas—and were helped by the UN High Commission for Refugees. Aid workers reported that one-fifth of children were malnourished and that malaria was rife.

These were the conditions—war, exodus, famine, and disease—in which the GPEI had to coordinate with those nongovernmental organizations still managing to function in the north of Mali in an effort to maintain polio immunity in the beleaguered population. In the measured language of their May 2012 bulletin, GPEI reported that the security situation in Mali made it difficult for vaccination teams to reach local populations in some areas. In the south the month's vaccination activities proceeded without incident, but in the north a "humanitarian corridor" was not yet functional, they said, "and this may complicate reaching children." In response to the emergency, GPEI scaled up its technical support to identify ways of making sure the entire country received vaccination coverage. Special measures were explored to reach populations in Mali and those displaced and on the move. One of these was the adding-on of oral polio vaccine to the existing program of humanitarian interventions in the Sahel region, the biogeographic zone of transition between the Sahara Desert and the Sudanian Savannas, literally the "shore" or "coast" of vegetation marking the southern boundary of the desert sands. Among those targeted were 63,000 refugee children in neighboring Mauritania and 56,000 in Burkina Faso.

The importance of this large, impoverished region for polio eradication lies in its relationship with northern Nigeria, where polio remains endemic. Unstable Mali and its neighbor states are part of the polio importation belt of West and Central Africa, which remains at high risk of re-infection, not least because nomadic peoples numbering 10 million journey through northern Nigeria and its neighboring states in ancient patterns of migration.

In August 2012 the GPEI reported twelve cases of polio compared with only three at the same point in 2011 in the Nigerian states of Zamfara and Sokoto. Operational difficulties—challenges in planning and implementing vaccination rounds, which prevented children from being reached—were blamed. The large population, spread across formidable terrain, had made it difficult to find every last child in round after round of vaccination, particularly with the moving target of nomadic communities. And even in the settled population, reaching every child in the remote villages, often inaccessible by road, had proved a daunting task. The climate of Zamfara is warm and tropical, with temperatures up to 40°C from March to May. The rainy season lasts from May until September, while the cold season holds sway from December to February. Winter brings a trade wind, the *Harmattan,* carrying the dust of the Sahara to the plains and uplands of northern Nigeria, drying the air and weakening the sun. In summer Sokoto City, in the dry Sahel surrounded by sandy savannah and isolated hills, is one of the hottest cities in the world. Time and again vaccination teams and their supervisors traveled long distances in every season. Keeping morale up and team members motivated proved a constant challenge.

But often the reason why children are missed in northern Nigeria is not the challenges of climate or desert or lack of roads. Sometimes it's much simpler: the children just aren't home when the vaccinator calls. In Zamfara and Sokoto, it is not unusual for children as young as three to be sent out to herd cattle. Others may be looked after at the playground by brothers or sisters themselves as young as five years old. Figures for Sokoto collected in the September 2011 GPEI vaccination campaign showed, for example, that 32 percent of children who were absent from the house were at the playground, 27 percent were in the fields, and 19 percent were away attending a social event with their mother or father.

The GPEI program strives to tackle these operational difficulties with fresh ideas. Microplanning strategies in Nigeria have been refined using lessons learned from India's successful eradication program. Mapping and oversight have been improved by the introduction of Global Positioning Systems and other mobile technologies. The "Nomad Project" is producing plans to reach nomadic populations with the vaccine and also with communications specifically tailored to these cultures and communities. Vaccinators who consistently reach a high percentage of children are specially rewarded, inspiring those on the ground to keep on doing their best. Thankfully, hardworking Nigerian polio personnel have had their loads lightened by a threefold increase in human resources in the region: in July and August 2012, 2,000 new staff were trained as part of a drive to reenergize, improve, and refresh the program at a crucial time.

Ever since October 2001, Afghanistan has been in the grip of war as Western forces and the Afghan United Front wage their campaign against the al-Qaeda terrorist organization and the Taliban. Traditionally, the armed conflict and the lack of security in southern provinces has hampered vaccination efforts, but surprisingly since 2012, the Afghan war is no longer the most important obstacle to success. Although 75 percent of Afghanistan's polio cases in the past decade have been in Helmand and Kandahar and the neighboring troubled provinces, vaccination campaigns in March 2012 reached more than 95 percent of children in the most challenging areas compared with only 70 percent at the beginning of 2011.

But even with this increased access, vaccination coverage has sadly declined. The latest difficulties, according to GPEI bulletins, lie not in operating in a war zone but in overcoming much more mundane problems. Children in small pockets across populated areas miss out on routine polio vaccination despite vaccinators adhering to standard practices. The Independent Monitoring Board for polio eradication has called these "polio sanctuaries." Data collected during campaigns indicates that 23 percent of unvaccinated children in the southern region of Afghanistan escape vaccination for social reasons that are part and parcel of traditional beliefs and practices.

In some communities, strong traditional beliefs surround how and when a child should be vaccinated. Children who are sick, sleeping, or newly born are not volunteered for vaccination when the teams arrive.

Vaccinators, though technically efficient, often lack training in how best to establish socially sensitive relationships with parents to find out just what's going on, and supervisors fail to spot the problem. Female vaccinators, who may be more easily accepted into households such as these, are more difficult to recruit. On top of this, the evidence suggests that because of shortcomings in planning, training, and supervision, some houses simply get missed.

To address these difficulties, the Afghan government has put together an action plan for greater focus on management and accountability, including the establishment of new teams at district level—a vital part of the larger Emergency Plan. To meet the challenge of villages in the war zone, the campaign recruited teams living within the community that can operate autonomously during lulls in hostilities. The GPEI points to successes in this most challenging land: 96 percent of Afghan children have received at least one vaccination against polio, and the campaign's goal is to reach every last child. In September 2012, President Hamid Karzai formally endorsed the national plan as the third nationwide immunization campaign in Afghanistan got underway.

The province of Sindh lies in the south of Pakistan, its wide plains watered by the Indus River. Some say its name comes from an Indo-Aryan belief that the Indus flows from the mouth of *Sinh-ka-bab,* the lion, but others trace the name of this flat, open land to the Sanskrit, *Sindhu,* meaning ocean. To the north lie Balochistan and the Punjab; eastward are the Indian states of Gujarat and Rajasthan; southward is the Arabian Sea. The historic town of Thatta, once a celebrated place of learning, art, and commerce, and a thriving port until the Indus altered its course, rises in the Sindhi plain some 60 miles west of Karachi. Here the visitor can marvel at the beauty of the Jamia Mosque, built by the Moghal Emperor Shah Jehan and adorned exquisitely with vivid blue glazed tiles. Not far from its 100 domes lies the vast Makli Necropolis, famed for its fine architecture, stone carvings, and glazed decoration.

Sanwan village lies nearby, in the melon-growing margins of the Indus, and here thirteen-month-old Sundar Lal's family has been farming melons for generations. In August 2012, the GPEI told the story of this little boy. The hot, dry air and sunshine of the fruiting season gives the musk melon, prized in the markets of Karachi and the Punjab, its sweetness, but it seems unlikely that Sundar will help with the harvest in years to come. In January 2012 he was diagnosed with polio. The events surrounding his infection illustrate the many operational challenges the GPEI must deal with in the field.

Sundar's life began in a camp for internally displaced persons (IDP) following the floods that hit the Hyderabad district in 2010. Despite this difficult beginning, his mother succeeded in getting him to the nearby clinic run by the WHO Expanded Program on Immunization (EPI) for his first infant vaccinations. But in areas like the IDP camp, with poor sanitation and the risks of malnutrition and diarrheal diseases, the oral polio vaccine needs to be administered at least four times to be sure of effectiveness. To achieve this, the campaign mounts SIAs—supplementary immunization activities—added on to children's routine vaccination dates. But an investigation into Sundar's case later found that vaccination teams assigned to his area did not follow up after Sundar was registered as "not available" during the vaccination round. They blamed this on team members' fears of stray dogs roaming the neighborhood.

But this was not the only failure: during the planning, three separate areas were considered as one for the purposes of microplanning, and the vaccinations were assigned to a single person. This doubled the number of children this worker was supposed to reach. As a result, Sundar was not picked up in supplementary immunization drives, leaving him vulnerable to polio. In response to Sundar's case and others like it, a surge in vaccination efforts by local staff and by WHO and UNICEF personnel was instituted. The aim was to raise the performance of vaccination teams, improve microplanning, and deliver good standards at all levels of operation. The new approach is paying off: there were only two more cases of polio after Sundar's in Sindh province in the first six months of 2012, compared with fourteen for the same period in 2011.

The lessons learned from success in India, once the world's largest exporter of wild poliovirus, are being used throughout the world. India is now exporting not disease but expert staff to support the remaining endemic countries. India also hosts visits from neighboring governments with the aim of sharing the best practices. These include methods to maximize coverage during vaccination rounds, media advocacy strategies, monitoring and evaluation formats, and ground-level tools to increase the effectiveness of front-line vaccinators and social mobilizers. In some areas of Pakistan, localized resistance stems from the persistent belief that the polio program is a Western plot to sterilize Muslims, or that the contents of the vaccine are not halal (that is, not produced in accordance with Islamic laws that govern products consumed by Muslims). One role of social mobilizers is to explain and dispel these myths about the vaccine. Mobilizers use passages from the Qur'an and the Hadith to remind people of the importance of the health of children in Islam. Pronouncements by highly respected Islamic clerics also help to demonstrate that polio vaccine is not anti-Islamic. Once persuaded, religious leaders then have the influence to convince families to vaccinate their children.

In areas that don't have access to television or radio, face-to-face meetings between social mobilizers and parents are used to enlighten and encourage those who might otherwise refuse vaccination for religious

reasons. Across more than 30 districts in Pakistan at high risk of refus-
als, one out of every two parents who initially declines vaccination for
their children are later convinced of its value, according to a UNICEF
survey. Thanks to the work of religious leaders, the total number of
polio cases in Balochistan, one of the four provinces of Pakistan, was
limited to three in the first nine months of 2012. All these measures are
intended to stop those individual tragedies, like Sundar's, which might
have been prevented.

Among China's 1.35 billion people, October 9, 2012, marked one
year since the last case of polio. In the year before, polio had arrived
from Pakistan, causing an outbreak in southern Xinjiang Uyghur Autonomous Region. This land of mountains, glaciers, forests, and grasslands, crossed by the ancient Silk Road, lies in the far northwest of
China, reaching up between Mongolia and Kazakhstan. It is the largest
province in the People's Republic and the homeland of the Uighur people, Turkic-speaking Muslims whose lands have been a part of China
since the eighteenth century. Before this outbreak, the whole of China
had been polio-free since 1999, with the last indigenous case occurring
in 1994.

As soon as the outbreak was detected, the Chinese government
mounted an exemplary response. Within hours of confirmation, on August 25, 2011, a public health emergency was declared. The minister and
vice-minister of health flew to the affected region to lead personally the
all-out response. Within days over 5 million doses of oral polio vaccine
had been airlifted to the province, and more than a half-million volunteers, health workers, and government officials had been mobilized to
support the immunization campaigns. The vaccinations were conducted
house to house but also at emergency vaccination posts set up at key
gathering places: bus and railway stations, schools, airports, bazaars,
city intersections, on motorways, at road check points, and in the remotest reaches of the province. Teachers, religious leaders, and community
figures were mobilized to spread the word about the vital importance of
immunization. TV, radio, and electronic media communications were
strengthened by several million banners and posters and by countless
leaflets. Stage plays were performed in village halls and squares, warning
of the dangers of infection and singing the praises of vaccination. Extensive monitoring systems were established that included the all-important
ear-, finger-, or arm-marking to identify those who had been immunized.

At the same time, the epidemiology of the virus was tracked to fa-
cilitate further targeting of the emergency response. More than 200,000
hospital records were reviewed, and almost 2 million households were
searched for suspected cases. Polio presents itself as acute flaccid paralysis, and samples from more than a thousand such cases were collected

and tested for virus to diagnose the cause. Every means of transport, including donkey cart, was used to deliver vaccine to the wildest reaches of the province. Altogether, five large-scale immunization campaigns were conducted in Xinjiang, and more than 43 million doses of oral polio vaccine were administered to populations of all ages. In total, the government of China allocated the equivalent of U.S. $55 million to the response effort.

In June 2012, to assess the overall quality of the outbreak response and determine the likelihood of ongoing transmission, China welcomed an international team of experts from the WHO and the U.S. Centers for Disease Control and Prevention. The team worked on the ground in Xinjiang and conducted a rigorous assessment of the effectiveness and sensitivity of the surveillance campaign. This international team concluded that it was "highly unlikely" that undetected wild poliovirus persisted in the Xinjiang Uyghur Autonomous Region. China's rigorous response had stopped the outbreak in record-time: it was just three months from the outbreak to the last case of polio. The likely consequence is that the Western Pacific Region will be able to maintain its polio-free certified status into the future.

The defeat of polio, when it happens, will be the second global eradication of a deadly disease afflicting humans. The greatest collaborative human achievements often take place in advanced nations with lavish funding and few limitations on resources through the efforts of elite teams in hothouse institutions. They are often motivated by intense international competition. Putting a man on the moon is an obvious example, as is the development of the internet and nuclear weapons.

But disease eradication is a different species of endeavor. Inevitably it takes place in remote and challenging regions of the world frequently troubled by armed conflict, repeatedly beset by drought, flood, and famine, and afflicted by poverty and political turbulence. The stories of smallpox and polio campaigns are full of such chapters. And the vast teams of people joined in the endeavor speak scores of different languages, subscribe to different values, worship different gods, come from almost every culture, and are united only in the global task. There are often huge setbacks, and no doubt there are huge inefficiencies. We know there are periods of great frustration and hopelessness. Certainly there are resource limitations: the WHO hoped for a budget of $2.6 billion for combating polio in 2010–2012 but says it received only half that amount. The Strategic Advisory Group of Experts on Immunization (SAGE) met in Geneva in November 2012 and expressed alarm at the funding shortfalls for GPEI at a time when eradication is in sight. And yet I feel we can be confident that in time the campaign will prevail.

24

THE MILKMAID AND
THE CUCKOO

"'It's a poor sort of memory that only works backwards,' the Queen remarked."

Lewis Carroll, *Through the Looking-Glass,
and What Alice Found There*, 1871

The library of Cambridge University ("the UL") was completed in 1933 to the grand design of Giles Gilbert Scott. It looks a lot like Bankside Power Station (now London's Tate Modern art gallery), which Scott designed fourteen years later. It's a sort of cathedral to books, and in the Rare Books Collection is a copy of Jenner's *Inquiry*. When I asked to see it, the whispering librarians produced it within a half hour. No white gloves were needed—it's a second edition bequeathed by Geoffrey Keynes, biographer, surgeon, physician, and literary bibliophile. Resting on protective cushions, the book's covers could hardly be more suitable: it's bound in *calfskin*. Inside the cover is a transferred label inscribed in Jenner's hand, and the book is dedicated to the king. Its pleasing, large-size print occupies a little over 70 pages, but the most arresting thing about seeing the original is the impact of the four colored plates. In Jenner's meticulous drawing of cowpox pustules, the hand of Sarah Nelmes extends, huge and unframed, across the empty page, her slender fingers open and relaxed, enhanced by sepia hatching and fine point-work whose crowning feature is the rose-colored wash around the blister just below her thumb. Jenner describes this redness, in the third plate (the arm of eight-year-old William Pead), as an efflorescent blush.

Loose inside the cover I found an article presented to Sir Geoffrey Keynes by its author, William Le Fanu, librarian to the Royal College of Surgeons. "Shall we agree with the critics," Le Fanu asks, "that Jenner was a dreamer who foisted on the world an unnecessary treatment, which he did not fully understand and which had been know for twenty years before? Or . . . do we stand with Pasteur and Lister, with Pirquet and Mellanby and say that Jenner's work was the most beneficent and fertile influence on the subsequent progress of medicine?"

When, in 1798, Edward Jenner set in train the events that have led, in a little over 200 years, to the first global eradications of disease, he kept a journal. In it he tells how he loves to roam the countryside of the Severn Vale and reflect upon the prospect of taking away from the world one of its greatest calamities. This he wisely balances with the prospect of personal independence and domestic harmony as sources of a happiness "so excessive that, in pursuing my favorite subject among the meadows, I have sometimes found myself in a kind of reverie."

I would like to join him on these dreamlike journeys and enjoy his company and conversation. I'd like to walk with him a mile or two, and pause beside the river, and find the time to talk of many things: of ships that sailed with orphan boys, of sealing wax that closed his vaccine quills, of kings who took his part and championed his cause (Lewis Carroll, by the way, wrote ardently on the value of vaccination). In the ancient shale beds that line the Severn, the fossils prized by Jenner still lie hidden. Here we could explore the distant past, but also the future. Jenner could not have foreseen the elimination of polio through vaccination because it simply wasn't known in his lifetime. There were no polio epidemics, and poliomyelitis wasn't recognized as a disease until Jakob Heine, a German orthopedist, described it in 1840.

What might impress him most is the story of the second historic vaccine to be derived from cows: BCG (Bacillus Calmette–Guérin) protects against tuberculosis, the insidious infection that robbed Jenner of his wife and eldest son. But BCG is only effective in temperate latitudes. It doesn't protect those who live in regions closer to the equator, perhaps because immune profiles are different in people long exposed to tropical parasites, and we need a better vaccine.

The countryside of south Gloucestershire is a storybook landscape: hills from a child's painting, secret valleys, waysides strewn with wild flowers, and forest softening the skyline. I'd like to ride with Jenner up the sudden Cotswold scarp in chase of an errant balloon; and at the ending of the headlong gallop, arrive at Kingscote and meet Catherine, soon to be his wife. I'd like to travel in a horse-drawn gig to the Fleece Inn, nestled in its wooded valley, and join in the convivial talk, pondering by candlelight, arguing by firelight, savoring a mutton supper, and musing our way toward some essential yet elusive truth. For Jenner's insight had to do not with the arcane abstractions of natural philosophy but with seeing, written in nature's web, secrets anyone might hope to understand. The fair complexion of the milkmaid and the hidden life of the cuckoo require an explanation.

This is a poet's country. To the south, silent valleys hoard untold histories and widen into broad green waves, opening at last onto the vales of great and lesser dairies where Thomas Hardy sent his milkmaid, Tess, out of her rural idyll and into the modern age. To the east, where Jenner went to school, the valleys narrow and grow deep—a winter country and a summer country without connection, according to poet Laurie Lee, who also went to school nearby: "The June air infected us with primitive hungers, grass seed and thistle down idled through the windows, we smelt the fields and were tormented by cuckoos." Lee's childhood valley was "a still green pool flooded with the honeyed tide of summer," where he sat on the floor like a "fat young cuckoo" while his sisters fed him currants.

All country people hearken to the cuckoo, the herald of ancestral spring, but Edward Jenner was only an amateur poet. Deeper down he was a scientist, and though the robin in his garden inspired him to verse, the cuckoo needed serious attention.

I'd like to climb the hillsides with the hawthorn in full flower and join his nephew Henry in his search for nests suitable for study. Up there we'd find them out at last, ingeniously hidden in the riot of spring growth, and delicately part the leaves to see the frightened bird still brooding or gaze at pristine, sky-blue eggs, fragile and yet faultless, each perfect and mysteriously closed upon itself. All this poised perfection will be shattered when the cuckoo chick breaks out. No one else was sufficiently assiduous to confirm Jenner's findings on the habits of the cuckoo until 1921, when the behavior of cuckoo chicks was finally photographed.

I'd like to join him in his duties as vaccine clerk to the world, surrounded by historic letters with newly broken seals, and marvel at his correspondents: Thomas Jefferson, Napoleon, the Tsar of Russia, and the Bourbon king of Spain. Jenner's discovery, and his campaign to tell the world, eventually affected the lives of countless millions. His vaccine reached to the farthest corners of the globe. But like many of the greatest stories, the essential cast of characters is small: Sarah Nelmes, the milkmaid who scratched her hand on a thorn and let the cowpox in; James Phipps, the gardener's boy; Hannah Excell, age eleven; Mary Pead, age five; Mary James, age six; Robert Jenner, age eleven months; and the other members of the small band of children Jenner immunized to protect them against smallpox. Other children also came to symbolize this great endeavor: Ann Dusthall, the first to be vaccinated on Indian soil, who became the primary source for vaccine snaking chain-wise through the whole subcontinent of India; and the foundling boy (his name unknown) who was the very last child bearing vaccination pustules as Francisco Xavier Balmis sailed into Caracas. This one boy's donation began the vaccine chain that spanned the continent of South America.

The Severn River seems to widen in midwinter; the farther shore at Berkeley turns into a vanishing horizon, the Cotswold scarp a shadow in the eastern sky. I'd like to sit with Jenner in his study late at night, encircled by the lamplight, listening to creaking timbers and the fire's soft hiss and crackle. In the last lateness of the country night, when the day's demands have faded and it's easier to slip the bonds of time, I'd like to tell him of some other names. Eugene Le Bar, who died aged 47, and Carmen Acosta, who died at 26 carrying her unborn baby—three among the last victims of smallpox in the United States of America. Saiban Bibi, age 30, who lived on the railway station in Karaminganj, the last person on Indian soil to suffer smallpox. Rahima Banu, age three, who lived on Bhola Island in the mouth of the Meghna River, the last person in Bangladesh to suffer smallpox and the last person in the world to be naturally infected with *Variola major*. Ali Maow Maalin of Somalia, the last person in human history to be naturally infected with smallpox.

There is another name I'd like to tell him, but I do not know it yet. It is the name of the last person in history to be naturally infected with polio. The chances are they're living now, in Pakistan, or Afghanistan, or in the heart of Africa. If not, the chances are they'll very soon be born. And sooner or later, thanks to vaccination and to the efforts of an indefatigable army—a team of many colors speaking scores of languages, bound by different values, raised with differing beliefs, drawn from different tribes and diverse cultures from every corner of the world—we will know that name.

ACKNOWLEDGMENTS

Heartfelt thanks to my agent, Peter Tallack, for helping shape and balance the scope and vision of this book from very early in the project and guiding it so successfully toward publication. I am extremely grateful to Luba Ostashevsky of Palgrave Macmillan for her wisdom, humor, and conviction in improving this narrative enormously and educating me in the process. I am likewise indebted to Laura Lancaster for her patience and her tireless improvements to the text. Thanks also to Bill Warhop for refining my style while retaining my English voice. I would like to express my gratitude to Carla Benton and Katherine Haigler for guiding me through the later stages. Tracey Lillis and Lauren Dwyer also deserve my warmest thanks. I remain extremely grateful to Lorraine Joneman and Claire Morrison of Palgrave Macmillan, to Helen Kennedy of the Jenner Museum, and to Dan Price of the British Society for Immunology. Particular thanks must go to Miriam Ward, Karen Baumgartner, Beth Bahls, David Rose, Roger Ireland, and Leonie Rhodes, for making possible the illustrations which enrich this book. Finally, I am delighted that the book benefits from the distinguished contribution of Sir Richard Sykes.

For their unfailing support, constant encouragement and sound advice I thank my wife Heather Rhodes and my daughters Chloe Rhodes and Leonie Rhodes.

BIBLIOGRAPHY

Alilbek, K. *Biohazard*. New York: Random House, 1999.

Allen, A. *Vaccine: The Controversial Story of Medicine's Greatest Lifesaver.* New York: W. W. Norton & Company, 2007.

Artenstein, A. W. "A Brief History of Influenza and Early Virology." In *Vaccines: A Biography,* edited by A. W. Artenstein, 191–205. New York: Springer Science, 2010.

Baron, J. *Life of Edward Jenner MD with Illustrations from His Doctrines and Selections from His Correspondence.* 2 vols. London: H. Colburn, 1827–1836.

Barret, A. T., and L. R. Stanberry. *Vaccines for Biodefense and Emerging and Neglected Diseases.* Waltham, MA: Academic Press, 2008.

Barry, J. M. *The Great Influenza: The Story of the Deadliest Pandemic in History.* New York: Penguin Books, 2004.

Baxby, D. *Jenner's Smallpox Vaccine: The Riddle of Vaccinia Virus and Its Origins.* London: Heinemann Educational Books, 1981.

Baxby, D. "Review of *Studies in Smallpox and Vaccination,* by William Hanna." *Reviews in Medical Virology* 12 (2002): 202–9.

Baxby, D. *Vaccination: Jenner's Legacy.* Berkeley: The Jenner Educational Trust, 1994.

Baxby. D. "Edward Jenner's Inquiry: A Bicentenary Analysis." *Vaccine* 17 (1999): 301–7.

Bazin, H. *The Eradication of Smallpox,* translated by Andrew and Glenise Morgan. London: Academic Press, 2000.

Beall, O. T., and R. H. Shryock. *Cotton Mather: First Significant Figure in American Medicine.* Baltimore: Johns Hopkins University Press, 1954.

Behring, E. A. *Serum Therapy in Therapeutics and Medical Science.* Nobel Lectures in Physiology or Medicine 1901–1921, Amsterdam: Elsevier Publishing Company, 1967.

Bender, G. *Great Moments in Medicine.* Detroit: Park Davis, 1961.

Biggs, J. T. *Leicester: Sanitation Versus Vaccination: Its Vital Statistics Compared with Those of Other Towns, the Army, Navy, Japan, and England and Wales.* London: The National Anti-Vaccination League, 1912.

Bodian, D. "Pathogenesis of Poliomyelitis." *American Journal of Public Health Nations Health* 42 (1952): 1388–1402.

Bowers, J. Z. "The Odyssey of Smallpox Vaccination." *Bulletin of Medical History* 55 (1981): 17–33.

Brock T. D. *Robert Koch: A Life in Medicine and Bacteriology.* Washington, D.C.: American Society for Microbiology, 1999.

Burnet. F. M. *Immunological Recognition of Self.* Nobel Lectures in Physiology or Medicine 1942–1962, Amsterdam: Elsevier Publishing Company, 1964.

Burnet. F. M. *The Clonal Selection Theory of Acquired Immunity.* Cambridge: Cambridge University Press, 1959.

Carrol, L. *Three Letters on Anti-vaccination, 1877.* Republished by The Lewis Carroll Society, London, 1976.

Chen, W., G. B. Patel, H. Yan et al. "Recent Advances in the Development of Novel Mucosal Adjuvants and Antigen Delivery Systems." *Human Vaccines* 6 (2010): 706–14.

Clyde, D. F., H. Most, V. C. McCarthy, and J. P. Vanderbert. "Immunization of Man Against Sporozoite-induced Falciparum Malaria." *The American Journal of the Medical Sciences* 226 (1973): 169–77.

Cohen, I. R. *Tending Adam's Garden: Evolving the Cognitive Immune Self.* London: Academic Press, 2000.

Cooke J. G., and F. Tahir. *Polio in Nigeria: The Race to Eradication.* Center for Strategic and International Studies, Global Health Policy Center report, 2012.

Crawford, D. H. *Deadly Companions: How Microbes Shaped Our History.* Oxford: Oxford University Press, 2009.

de Kruif, P. *Microbe Hunters.* Boston: Houghton Mifflin Harcourt, 2002. First published 1926 by Harcourt Brace Jovanovich.

Defoe, D. *A Journal of the Plague Year.* London: Penguin Books Ltd., 2003.

Dubos, R. *Pasteur and Modern Science.* Washington, D.C.: American Society for Microbiology, 1998.

Dubos, R., and J. Dubos. *The White Plague: Tuberculosis, Man and Society.* New Brunswick: Rutgers University Press, Revised edition 1987.

Fenner, F., D. A. Henderson, J. Arita, Z. Jezek, and I. D. Ladynyi, *Smallpox and Its Eradication.* Geneva: World Health Organization, 1988.

Fisher, R. B. *Edward Jenner: A Biography.* London: Andre Deutsch, 1991.

Foege, W. H. *House on Fire: The Fight to Eradicate Smallpox.* Berkeley: University of California Press, 2011.

Foege, W. H., J. D. Millar, and D. A. Henderson. "Smallpox Eradication in West and Central Africa." *Bulletin of the World Health Organization* 52 (1975): 209–22.

Fosbroke, T. D. *Berkeley Manuscripts.* London: John Nichols, 1821.

Fracastoro, G. *De contagionibus et contagiosis morbis et earum curatione (On Contagion and Contagious Diseases),* Venice, 1555.

Franco-Paredes, C., L. Lammoglia, and J. I. Santos-Preciado. *The Spanish Royal Philanthropic Expeditions to Bring Smallpox Vaccination to the New World and Asia in the 19th Century. Clinical Infectious Diseases* 41 (2005): 1285–89.

Glynn, I., and J. Glynn. *The Life and Death of Smallpox.* London: Profile Books, 2004.

Gould, A. *Summer Plague: Polio and Its Survivors.* New Haven: Yale University Press, 1995.

Heller, J. *The Vaccine Narrative.* Nashville: Vanderbilt University Press, 2008.

Henderson, D. A. *Smallpox: The Death of a Disease.* Amherst, NY: Prometheus Books, 2009.

Henderson, D. A. "Smallpox Eradication—the Final Battle." *Journal of Clinical Pathology* 28 (1975): 843–49.

Hoffman, S. L., L. M. L. Goh, T. C. Luke et al. "Protection of Humans Against Malaria by Immunization with Radiation-attenuated *Plasmodium falciparum* Sporozoites." *Journal of the Royal Society of Tropical Medicine and Hygiene* 185 (2002): 1155–64.

Hunter, J. *Letters from the Past from John Hunter to Edward Jenner.* Edited by E. H. Cornelius and A. J. Harding Rains. London: Royal College of Surgeons of England, 1976.

Institute of Medicine Report. *Adverse Events Associated with Childhood Vaccines: Evidence Bearing on Causality.* Washington, D.C.: National Academy Press, 1994.

Institute of Medicine Report. *DPT Vaccine and Chronic Nervous System Dysfunction: A New Analysis.* Washington, D.C.: National Academy Press, 1994.

Institute of Medicine Report. *Immunization Safety Review: Influenza Vaccines and Neurological Complications.* Washington, D.C.: The National Academy Press, 2003.

Institute of Medicine Report: *Immunization Safety Review: Measles-Mumps-Rubella Vaccine and Autism.* Washington, D.C.: National Academy Press, 2001.

Institute of Medicine Report: *Vaccine Safety Research, Data Access, and Public Trust.* Washington, D.C.: National Academy Press, 2005.

Institute of Medicine Report. *Immunization Safety Review: SV40 Contamination of Polio Vaccine and Cancer.* Washington, D.C.: The National Academy Press, 2002.

Janeway, Jr., C. A. "Approaching the Asymptote? Evolution and Revolution in Immunology." *Cold Spring Harbor Symposium on Quantitative Biology* 54 (1989): 1–13.

"Jenner Centenary Abroad." *British Medical Journal* 1:1291 (1896).

Jenner, E. *An Inquiry into the Causes and Effect of the Variolae Vaccinae—A Disease Discovered in Some of the Western Counties of England, Particularly Gloucestershire, and Known by the Name of the Cow Pox.* London: Sampson Low, 1798.

Jenner, E. *Collections of Poems and Songs,* 1794. MS 3017 9vol 1) and 3016 (vol 2). Wellcome Library Archives.

Jenner, E. *Further Observations on the Variolae Vaccinae or Cow-Pox.* London: Sampson Low, 1799.

Jenner, E. "Observations of the Natural History of the Cuckoo. By Mr. Edward Jenner, in a letter to John Hunter, Esq. FRS." *Philosophical Transactions of the Royal Society* 78 (1788): 210–37.

Kendall, M. D. *Dying to Live: How Our Bodies Fight Disease.* Cambridge: Cambridge University Press, 1998.

Kluger, J. *Splendid Solution: Jonas Salk and the Conquest of Polio.* New York: G.P. Putnam's Sons, 2004.

Koch, H. H. R. *Investigations into the Etiology of Traumatic Infective Diseases.* London: The New Sydenham Society, 1880.

Koch, H. H. R. *The Current State of the Struggle against Tuberculosis.* Nobel Lectures in Physiology or Medicine 1901–1921, Amsterdam: Elsevier Publishing Company, 1967.

Koprowski, H. "Historical Aspects of the Development of Live Virus Vaccine in Poliomyelitis." *British Medical Journal* 2 (1960): 85–91.

Lambert, S. M., and H. Markel. "Making History: Thomas Francis, Jr, MD, and the 1954 Salk Poliomyelitis Vaccine Field Trial." *Archives of Pediatrics & Adolescent Medicine* 154 (2000): 512–17.

Lancet editorial. "The Tragedy with BCG Vaccine at Lübeck." *Lancet* 215 (1930): 1137–38.

Le Fanu, W. R. *A Bio-Bibliography of Edward Jenner 1749–1823*. London: Harvey and Blythe, 1951.

Ling Lin, P., J. Dietrich, E. Tan et al. "The Multistage Vaccine H56 Boosts the Effects of BCG to Protect Cynomolgus Macaques Against Active Tuberculosis and Reactivation of Latent *Mycobacterium Tuberculosis* Infection." *Journal of Clinical Investigation* 122 (2012): 303–14.

McNeill, W. H. *Plagues and People*. Oxford: Blackwell, 1977.

McVail, J. C. "Cow-pox and Small-pox: Jenner, Woodville and Pearson." *British Medical Journal* 1 (1896): 1271–76.

Meldrum, M. "'A Calculated Risk': The Salk Polio Vaccine Field Trials of 1954." *British Medical Journal* 317 (1998): 1233.

Melville, L. *Lady Mary Wortley Montagu: Her Life and Letters (1689–1762)*. Stockbridge, MA: Hard Press, 2006.

Mitchison, N. A. "Passive Transfer of Transplantation Immunity." *Nature* 171 (1953): 267–68.

Mitchison, N. A. "T-cell–B-cell Cooperation." *Nature Reviews Immunology* 4 (April 2004): 308–12.

Nathanson N. "The Pathogenesis of Poliomyelitis: What We Don't Know." *Advances in Virus Research* 71 (2008): 1–50.

Newspaper report of the Linnemanstons family polio deaths in 1952. "Vaccine Too Late in this Case: Four in Family Dead." *Milwaukee Journal*, April 13, (1955): 18.

Obregón, R. K., Chitnis, C. Morry et al. "Achieving Polio Eradication: A Review of Health Communication Evidence and Lessons Learned in India and Pakistan." *Bulletin of the World Health Organization* 87, 2009: 624–630.

Oldstone, M. B. A. *Viruses, Plagues, and History: Past, Present and Future*. Rev. ed. Oxford: Oxford University Press, 2009.

Oshinsky, D. M. *Polio: An American Story*. Oxford: Oxford University Press, 2005.

Packard, R. M. *The Making of a Tropical Disease: A Short History of Malaria*. Baltimore: Johns Hopkins University Press, 2011.

Paul, W. E., ed. *Fundamental Immunology*. 7th rev. ed. Philadelphia: Lippincott Williams and Wilkins, 2012.

Pearson, G. C. *An Inquiry Concerning the History of the Cow Pox. Principally with a View to Supersede and Extinguish the Small Pox*. London: J. Johnson, 1798.

Porter, R. *The Greatest Benefit to Mankind: A Medical History of Humanity*. New York: W. W. Norton & Company, 1997.

Powell, A. "John Enders' Breakthrough Led to Polio Vaccine." *Harvard University Gazette*, October 8, 1998.

Preston, R. *The Demon in the Freezer*. New York: Random House, 2002.

Rappuoli, R. and A. Aderem. "A 2020 Vision for Vaccines Against HIV, Tuberculosis and Malaria." *Nature* 473 (2011): 463–69.

Reid A. H., J. K. Taubenberger, and T. G. Fanning. "The 1918 Spanish Influenza: Integrating History and Biology." *Microbes and Infection* 3 (2001): 81–87.

Renne, E. P. *The Politics of Polio in Northern Nigeria.* Bloomington: Indiana University Press, 2010.

Rerks-Ngarm, S., P. Pitisuttithum, S. Nitayaphan et al. "Vaccination with AL-VAC and AIDSVAX to Prevent HIV-1 Infection in Thailand." *New England Journal of Medicine* 361 (2009): 2209–20.

Rhazes (Abu Bakr Mohammad Ibn Zakariya al-Razi). *A Treatise on the Smallpox and Measles,* translated by W. A. Greenhill. Baltimore: Williams & Wilkins, 1939. First published 1848 by Sydenham Society, London.

Desowitz, R. *The Thorn in the Starfish.* New York: W. W. Norton & Company, 1987.

Roth, P. *Nemesis.* London: Jonathan Cape, 2010.

Royal College of Physicians. *Report of the Royal College of Physicians of London on Vaccination.* London: Society of the Poor, 1806. Reprint by L. Hansard for Longman, Hurst, Rees & Orme, London, 1807.

RTS,S Clinical Trials Partnership. "A Phase 3 Trial of RTS,S/AS01 Malaria Vaccine in African Infants." *New England Journal of Medicine* 367 (2012): 2284–95,

Ryan, F. *Tuberculosis: The Greatest Story Never Told—The Search for the Cure and the New Global Threat.* Sheffield: FPR-Books Ltd., 1992.

Sabin, A. B., and L. R. Boulger. "History of Sabin Attenuated Poliovirus Oral Live Vaccine Strains." *Journal of Biological Standardization* 1 (1973): 115–18.

Salk, J. E., U. Krech, and J. S. Younger. "Formaldehyde Treatment and Safety Testing of Experimental Poliomyelitis Vaccines." *American Journal of Public Health* 44 (1954): 563–70.

Sato, Y., and H. Sato. "Development of Acellular Pertussis Vaccines." *Biologicals* 27 (1999): 61–69.

Silver. J. *Post-Polio Syndrome: A Guide for Polio Survivors and Their Families.* New Haven: Yale University Press, 2001.

Silverstein, A. M. *A History of Immunology.* 2nd Edition. London: Academic Press, 2009.

Sloane, H., and T. Bart. "An Account of Inoculation by Sir Hans Sloane, Bart. Given to Mr. Ranby, to Be Published, Anno 1736." *Philosophical Transactions of the Royal Society* 49 (1755): 516–20.

Stöhr, K., M. P. Kieny, and D. Wood. "Influenza Pandemic Vaccines: How to Ensure a Low-cost, Low-dose Option." *Nature Reviews Microbiology* 4 (2006), 565–66.

Sykes, R. B. *New Medicines, the Practice of Medicine, and Public Policy.* London: The Nuffield Trust, The Stationery Office, 2000.

Thera, M. A., and C. V. Plowe. "Vaccines for Malaria: How Close Are We?" *Annual Review of Medicine* 63 (2011): 345–57.

Ulmer, J. B., B. Wahren, and M. A. Lui. "Gene-based Vaccines: Recent Technical and Clinical Advances." *Trends in Molecular Medicine* 12 (2006): 216–22.

VanBusirk, K. M., M. T. O'Neill, P. De La Vega et al. "Preerythrocyte, Live-attenuated *Plasmodium falciparum* Vaccine Candidate by Design." *Proceedings of the National Academy of Sciences* 106 (2009): 13004–9.

Virchow, R. L. K. *Cellularpathologie in Ihrer Begründung auf Physiologische und Pathologische Gewebelehre* [Cellular pathology as based upon physiological and pathological histology: twenty lectures delivered in the Pathological Institute of Berlin during the months of February, March, and April, 1858]. London: John Churchill, 1860.

Voltaire, François Marie Arouet de. *Letters on the English (Letter XI—On Inoculation)*. Edited by C. W. Eliot. The Harvard Classics. New York: P. F. Collier & Son, 1909–14.

Wakefield, A. J., S. H. Murch, A. Anthony et al. "Ileal-Lymphoid-Nodular-Hyperplasia, Non-specific Colitis, and Pervasive Developmental Disorder in Children." *Lancet* 351 (1998): 637–41.

Waltzer Leavitt, J. "Be Safe Be Sure: New York City's Experience with Epidemic Smallpox." In *Hives of Sickness: Public Health and Epidemics in New York City,* edited by D. Rosner, 95–113. New York: Museum of the City of New York, 1995.

Whiteside, A. *HIV/AIDS: A Very Short Introduction.* Oxford: Oxford University Press, 2008.

Williams, G. *Angel of Death: The Story of Smallpox.* London: Palgrave Macmillan, 2010.

Williams, G. *Paralyzed with Fear: The Story of Polio.* London: Palgrave Macmillan, 2013.

INDEX

Abubakar III, Muhammadu Sa'ad, 203
Acosta, Carmen, 153, 218
Acosta, Ismael, 152–3
Adams, John, 28
adaptive immunity, 2–3, 139–44, 195
adjuvants, 124–6, 129, 141, 150, 187, 194, 196
Advanced Research Project Agency (ARPA), 191–2
Afghanistan, 4, 57, 176, 199, 205, 207, 210–11, 218
Africa:
  HIV in, 98, 194
  malaria in, 196, 198
  polio in, 137, 199–209, 218
  smallpox in, 3, 52, 69, 152, 155–6, 159, 168–71, 191–2
  Spanish flu in, 147
  tuberculosis in, 98
  variolation in, 14, 57
  See also Ethiopia; Kenya; Mali; Nigeria; Somalia; South Africa
Agee, James, 55, 101
AIDS. See human immunodeficiency virus (HIV)
Al Rhazi, 5, 9, 11, 55
Albert, Prince, 90
Alibek, Kanatjan, 173, 176
Allen, Arthur, 179
Alston, Charles Henry, 111
Anderson, Donald, 138
Anderson, Elizabeth Garret, 66
anthrax, 81–2, 85–6, 89, 174
antibiotics:
  bacterial pneumonia and, 148
  future of, xi
  meningitis and, 202

penicillin, 112, 117, 148, 154
  resistance to, xi
  risk to benefit ratio for, xi
  streptomycin, 98
  trovaflaxin, 202
  yaws and, 156
antibody, discovery of, 2, 82–3
antigens, 195–6
antivaccination movements, 66–8, 90, 154, 179–90. See also vaccination: debate
Athens, plague of, ix, 8
attenuated vaccines, 91, 96, 106, 116, 119, 123–4, 126, 132–7, 151, 177, 183, 189, 194–5, 197
atypical measles syndrome, 183
Autism Spectrum Disorder, 188–9

Bacillus influenzae, 144. See also influenza
bacteria:
  antibody and, 83
  characteristics of, 79–87, 186
  compared with viruses, ix, 81
  discovery of, 79–80
  and infections at vaccination sites, 65
  and theory of spontaneous generation, 80–1
  See also individual bacterial infections
Baker, John, 42
Balmis, Francisco Xavier, 50–1, 57, 218
Bangladesh, 3, 165–8, 218
Banks, Joseph, 24, 38, 50
Banu, Rahima, 168, 218
Baron, John, 25, 30, 48, 52, 74

Baumgarten, Paul, 94
Baxby, Derrick, 177
BCG (Bacillus Calmette Guérin)
    vaccine, 95–99, 139, 216
Behring, Emil von, 82–3, 86, 89, 91
Beijerinck, Martinus, 81
Bell, Joseph A., 126–7
Bell, Thomas, 120
Bennett, Arnold, 143
Berkeley, Earl of, 23, 25, 37–8, 44
Berners-Lee, Tim, 192
Bibi, Saiban, 166, 218
Biggs, J. F., 67
Bill & Melinda Gates Foundation,
    192–3, 198
biochemistry, coinage of the term, 91
Biologics Act, 181
Biopreparat, 174–6
bioweapons, 4, 174–6
Birch, John, 41
Blair, William, 72
Bodian, David, 113–17, 124, 127–8
Boerhaave, Herman, 80
Boylston, Zabdiel, 19
Breinl, Friedrich, 87
Brodie, Maurice, 106–7
bubonic plague (Black Death), 6–7,
    13, 16, 85
Burnet, Frank Macfarlane, 2, 87, 145

Calmette, Albert, 95
Cantor, Eddie, 105–6
Caroline of Ansbach, Princess of
    Wales, 16–19
Carro, Jean de, 50
Carroll, Lewis, 215–16
Caverly, Charles, 102
Charles IV, King of Spain, 50
chickenpox, 7, 58, 81, 113, 140,
    152–3, 161, 166–7, 172
    chickenpox vaccine, 182, 185
China:
    deaths from infectious disease in,
        193
    Nixon's visit to, 163
    polio in, 213–14
    smallpox in, 57, 152, 175
    Spanish flu in, 147
    variolation in, 14, 18, 57
Chiswell, Sarah, 15
cholera, ix, 81, 85, 89, 91, 93, 106,
    112, 114, 116, 168, 201

Chumakov, Mikhail, 134
Clare, John, 23
Clark, James, 93
Clinch, John, 52
Cline, Henry, 24, 39
clonal selection theory of immunity,
    87, 145
Cohen, Joe, 198
Cohn, Ferdinand, 86
Cold War, 3, 127, 134, 154, 163,
    174–5, 191
Comite Central de Vaccine, 49
complement defense system, 143–4
consumption, 91–3. See also
    tuberculosis
Cook, James, 24, 38
Cortés, Hernando, 8
cowpox, 24, 30–5, 37–43, 49–50, 52,
    61, 65, 88, 95, 142, 177–8, 215,
    217
Cox, Harold, 135
Cross, John, 58
Cruikshank, Isaac, 48
cytomegalovirus, 143

Dan Maraya Jos, 204
Davaine, Casimir, 85
Defoe, Daniel, 6
Dimsdale, Thomas, 21, 44, 89
diphtheria, 2–3, 82–3, 85–6, 89, 106,
    114–15, 180–1, 186–7, 193
distemper, 144
DNA, 2, 32, 87, 140, 175, 177, 185,
    194
drift and shift, 144–5
DTaP vaccine, 186–7
DTP vaccine, 186–7
Dunning, Richard, 39, 57
Dusthall, Ann, 50, 217–18

ebola virus, xii
Eddy, Bernice, 182
Edmonston, Henry, 41
Ehrlich, Paul, 83, 86
Elgin, Lord, 50
Enders, John, 113, 116–17, 119,
    124–5, 132–3
epilepsy, 179
Ethiopia, 168–71, 201

Farr, William, 59
Fewster, John, 25, 30–1

Flexner, Simon, 103, 106–8, 113
flu. *See* influenza
Foege, William, 151, 157–8, 161–4, 167
formaldehyde, 91, 106, 119, 123, 150, 187
Fosbroke, Thomas Dudley, 21, 49
Fracastoro, Girolamo, 77–8
Francis Jr., Thomas, 110, 117–19, 124, 127–32, 145
Franco-Prussian War, 63, 85–6
Friedrich Wilhelm III, King of Prussia, 52

Gardner, Edward, 25, 33, 42
Gates Foundation. *See* Bill & Melinda Gates Foundation
Gillray, James, 41
GlaxoSmithKline, 185, 198
GlaxoWellcome, x
Global Alliance for Vaccines and Immunization (GAVI), 193
Global Polio Eradication Initiative (GPEI), 198–9, 202, 207–11, 214
Gorbachev, Mikhail, 175
Greuze, Jean-Baptiste, 30
Guérin, Camille, 95
Guillain–Barré syndrome, 184

H1N1 virus, 148–9
Hammond, William, 132
Hancock, John, 28
Hardy, Thomas, 71, 75, 217
Harrison, John, 44
Harvey, William, 29, 31
Haurowitz, Felix, 87
Haygarth, John, 52
Henderson, Donald, 155, 159–60, 165–6, 176
hepatitis A vaccine, 182
hepatitis B and serum-associated jaundice, 61, 181
hepatitis B vaccine, 3, 182–3, 185, 193, 197
Herberden, William, 27
herpes virus, 143, 176
Hilleman, Maurice, 182–5, 188
Home, Everard, 24, 38
horsepox, 43, 177
Horstmann, Dorothy, 114, 135
Howe, Howard, 114, 117, 125, 127

Hultin, Johan, 148–9
human immunodeficiency virus (HIV), 202
    and depletion of T-cells, 87–8
    and prevalence of TB, 98
    research to prevent, 1, 4, 193–6
    transmission of, xii
humors, 11, 83–6
Hunter, John, 24–8, 30–1, 38–9
Hunter, William, 24

immune system:
    and adaptive immunity, 2–3, 139–44, 195
    DTP vaccine and, 186–7
    hepatitis B vaccine and, 185, 197–8
    HIV/AIDS and, xii, 194
    influenza, ix, 150
    and instructive theories of immunity, 86–8, 139
    live vaccines' effect on, 119, 151, 189, 194–5
    polio vaccine and, 107, 134, 136–7, 206
    and serum sickness, 181
    smallpox vaccine and, 65, 176
    TB and, 92, 2, 98–9
immunization. *See* vaccination; variolation
immunology, x, 22, 83–8, 139, 141, 145
India, 3, 159–66, 199, 207, 212
Indonesia, 3, 50, 56, 155–6, 159
infant diarrhea. *See* rotavirus
infantile paralysis. *See* polio (poliomyelitis):
infectiousness, xii, 3–4, 8, 32, 34, 61–2, 80–2, 143–5, 193–5
influenza:
    and adaptive immune system, 3
    and drift and shift, 144–5
    H1N1, 148–9
    history and naming of, 144–5
    Spanish flu, 3, 145–9, 184
    types of, 149
influenza vaccine, 3–4, 96, 117, 119–20, 145, 149–51, 182, 184, 189
Information Age, 191–9
Information Processing Techniques Office, 191

inoculation, 15. *See also* vaccination; variolation

Ivanovsky, Dimitri, 81

Janeway, Charles, 141
Japan, 57–8, 187
Jefferson, Thomas, 53, 154, 217
Jenner, Edward:
  awards, honors, and statues for, 44–50, 90
  on Balmis, 51
  birth and childhood of, 23–4
  death and burial of, 74–5
  and death of son (Edward), 73, 216
  and death of wife (Catherine), 73, 92, 216
  as director of National Vaccine Institute, 48
  discovery of vaccination by, 1, 3, 37–45, 88, 142, 151, 193–4
  early career as country doctor, 24–30
  on eradication of smallpox, 55, 57–8, 63, 152, 166, 207
  first cowpox experiment of, 30–5
  Further Observations on the Variolae Vaccinae or Cow Pox, 40, 42
  influence on Pasteur, 81–2
  Inquiry into the Causes and Effect of the Variolae Vaccinae, An, 37–44, 48, 52, 178, 192, 215
  Jefferson's letter to, 53, 154, 217
  last visit to London by, 71–3
  Le Fanu on, 216
  marriage of, 27
  mistaken belief in lifelong protection, 59
  as observer, 31, 217
  as a poet, 25, 217
  typhoid fever contracted by, 31
  use of variolation by, 21, 23, 30
  wife of, 25–7
Jennerian vaccination, 48, 57, 59, 62, 65–6, 89–92, 119, 142, 151–2, 156, 177
Jerne, Neils, 87
Jervis, George, 115
Jesty, Benjamin, 34–5, 44
Jezek, Zdenck, 171

Jonathon, Goodluck, 203

Keats, John, 89, 92–3
Kempe, C. Henry, 183–4
Kenya, 169, 171, 198, 199
Keynes, Geoffrey, 215–16
Ki-moon, Ban, 199
Kirkpatrick, Bill, 124
Kirkpatrick, James, 78–9
Koch, Robert, x, 85–6, 91–2, 94–7, 144
Kolmer, John A., 106–7
Koprowski, Hilary, 115–16, 119, 123, 125, 133–5, 182, 195

L'Estrange, Roger, 13
Laidlaw, Patrick, 144
Landsteiner, Karl, 102–3, 113
Laurence, D. H., 191
Le Bar, Eugene, 152, 218
Le Fanu, William, 216
Lee, Laurie, 217
Legionnaires' disease, 184
Leibnitz, Gottfried, 18
Leicester Method, 67, 154–5, 158
leprosy, ix, 85
Lettsome, John Coakley, 43, 52, 73
life expectancy, ix, 6, 148
lipopolysaccharides, 186–7
Louis XV, King, 13, 20–1
Louis XVI, King, 21
Lucretius, 37
Ludlow, Daniel, 24
lymphocytes, 87–8, 140–2, 195

Maalin, Ali Maow, 172, 218
Macaulay, Thomas, 7, 11
Maitland, Charles, 15–18
malaria, 1, 4, 154–5, 172, 192–8, 201, 208
Malenkov, Georgi, 127
Mali, 208–9
Mann, Thomas, 97
March of Dimes, 2, 105–6, 110–11, 115, 128, 133, 138, 182, 192
Marshall, Joseph, 51
Marshall, William Calder, 90
Marston, J. F., 58
Mather, Cotton, 18–19, 78
McGregor, Ewan, 204–5
McLuhan, Herbert Marshall, 192

measles, xi, 3, 8–9, 96, 144, 151,
    156–7, 180, 182–3, 188, 193,
    205. *See also* rubella
Medawar, Peter, 87
Medley, Samuel, 73
Meister, Joseph, 82
meningitis, 103, 202
mercury, 188–9
Metchnikoff, Elie, 84–5, 139
Metchnikoff, Olga, 84
MHC (major histocompatibility
    complex) molecules, 140–3, 194
Millard, Killick, 65, 67
Mitchell, Albert, 145
MMR vaccine, xi, 188
Mohniké, Otto, 57
Montagu, Edward Wortley, 24
Montagu, Lady Mary Wortley,
    13–15, 19–20, 50, 170
Morgan, Isabel, 115
Mosley, Benjamin, 41
Mukherjee, Siddhartha, 91–2
Mycobacterium tuberculosis, 92, 95.
    See also tuberculosis (TB)

Nabokov, Vladimir, 109
Napoleon Bonaparte, 49–51, 72, 217
National Childhood Vaccine Injury
    Act, 186
National Foundation for Infantile
    Paralysis, 105, 110–13, 116–18,
    120, 123–6, 128, 131, 134, 136
National Immunization Days, 204
National Science Foundation (NSF),
    192
National Vaccine Institute, 47–8
Negri, Adelchi, 81
New Medicines, xi
Newell, Thomas, 49
Newgate Prison "Royal Experiment,"
    16–18
Newton, Isaac, 12, 18, 29
Nigeria, 4, 156–8, 16, 193, 199–204,
    207, 209–10
Nixon, Richard, 163, 174
Nomad Project, 210
Norton, Thomas, 115

O'Connor, Basil, 104–5, 110–12,
    117–8, 125–7, 130, 132
O'Doherty, Brian, 123

Obama, Barack, 174
Obasanjo, Olusegun, 201, 203
Olinsky, Peter, 103
Oshinsky, David, 120
Owen, Wilfred, 146

Pakistan, 4, 57, 155, 159, 167, 193,
    199, 205, 207, 211–13, 218
papilloma (wart) viruses, 143–4
Park, William H., 106
Parry, Caleb, 27, 38
Parton, James, 120
passaging, 95, 115, 134, 181
Pasteur, Louis, x, 77, 80–2, 85–6, 89,
    91, 112, 116, 119, 216
Paul, John, 116–17, 124, 127, 135
Pauling, Linus, 87
Pearson, Alexander, 57
Pearson, George, 39–41, 44, 48–9
peer review, 37, 192
penicillin, ix, 112, 117, 148, 154
pertussis vaccine, 186–7. *See also*
    whooping cough
Pfeiffer, Richard, 85, 144
Phipps, James, 31, 34, 38, 74, 88,
    142, 217
Pierrepont, Mary. *See* Montagu, Lady
    Mary Wortley
Piper, David, 79
Pius VII, Pope, 52
Plasmodium, 196–7. *See also* malaria
pneumonia, 3, 6, 85, 148, 193
polio (poliomyelitis):
    abortive, 137
    in Afghanistan, 4, 199, 205, 207,
        210–11, 218
    Brunhilde strain of, 120
    bulbar, 117, 137
    in China, 213–14
    campaign to eradicate, xi, 1, 3–4
    Franklin D. Roosevelt as victim of,
        103–5, 138
    history of, 101–12, 137–8
    incidence of, 4, 111–12, 117, 125,
        133, 136
    in India, 199, 207, 212
    Lansing strain of, 113, 115, 120
    Leon strain of, 120
    Mahoney strain of, 114, 120, 123,
        125, 128, 132–4
    in Mali, 208–9

in Nigeria, 4, 199–204, 207,
    209–10
nonparalytic, 137
in Pakistan, 4, 199, 205, 207,
    211–13, 218
and post-polio syndrome, 137–8
progression of disease, 2, 117,
    136–7
in Tajikistan, 206–7
transmission of, 102–3, 107,
    113–14, 136–7
types of, 107, 114–15, 120
See also March of Dimes; National
    Foundation for Infantile Paralysis
polio vaccines:
contamination of, 40, 182, 203
early testing of, 106–8
Letchworth Village trials, 115–16
Sabin vaccine, 133–8, 151, 182–3,
    195, 203, 206
Salk vaccine, 123–38, 183
poliomyelitis. See polio
    (poliomyelitis):
Pope, Alexander, 14
Popper, Erwin, 102
post-polio syndrome, 137–8
Priestly, Joseph, 67
public health, x, 4
    chickenpox and, 185
    in China, 213
    hepatitis B and, 185
    HIV/AIDS and, 194
    in Nigeria, 201
    polio and, 102–3, 106, 132, 208,
        213
    in Puerto Rico, 60
    smallpox and, 161–2, 169
Pylarini, Jacobo, 18

quarantine, 12, 68, 91, 93, 102, 111,
    154–5, 171

rabies, 82, 89, 114
Ramon, Gaston, 91
rational empiricism, 183
Rhodes, John, x–xi
Ring, John, 43
Rivers, Thomas, 112, 116, 124–5,
    127
Robbins, Frederick, 113
Roosevelt, Eleanor, 104–5

Roosevelt, Franklin Delano, 103–6,
    110–11, 138
Ross, Ronald, 194
Rossignol, Hippolyte, 82
rotavirus, 3, 96, 180, 185, 193
Roth, Phillip, 109–10
"Royal Experiment" (Newgate
    Prison), 16–18
Roux, Émile, 83
Rowley, Thomas, 41
Royal Jennerian Institution, 47
Royal Society, 16–18, 26–7, 37–8, 50,
    74, 79
rubella, 3, 96, 188, 193
Rutter, William, 185

Sabin, Albert, 103, 116–7
    polio vaccine of, 133–8, 151,
        182–3, 195, 203, 206
    and rivalry with Jonas Salk,
        119–20, 124–8, 132–8, 183
    work on poliovirus point of entry,
        113–14
Sacco, Luigi, 50–2
Salk, Jonas:
    birth and childhood of, 118
    education of, 118–19
    as founder of Salk Institute, 136
    as investigator for research funding,
        114, 116–18
    mentored by Thomas Francis Jr.,
        119, 127, 145
    at Pittsburgh Medical School,
        120–1
    polio vaccine of, 125–38, 183
    and rivalry with Albert Sabin,
        119–20, 124–8, 132–8, 183
    vaccine trials of, 2–3, 125–33, 182
    work on polio subtypes, 117, 120
Salmonella typhi, 31. See also typhoid
    fever
Sato, Yuji, 187
Scheele, Leonard, 132
Scott, Giles Gilbert, 215
serendipity, 112–13
Severn, Joseph, 93
Shakespeare, William: Romeo and
    Juliet, 78
Shaw, George Bernard, 66
Shekarau, Ibrahim, 202–3
Shibasaburo, Kitasato, 57, 82, 86

Shope, Richard, 144
Sloane, Sir Hans, 16–18
Smadel, Joseph, 125, 132
smallpox:
  in Africa, 3, 159–60, 168–71, 201, 218
  in Bangladesh, 3, 165, 167–8, 218
  and complement defense system, 143–4
  in England, 12–21, 160–9
  eradication of, xi, 1, 3–4, 47- 53, 55–63, 152–6, 159–63, 167–77, 183, 193, 195, 216
  in Ethiopia, 168–71, 201
  Hindu goddess of, 56–7
  history of, ix–x, 7–9, 11–13
  in India, 3, 159–66
  in Indonesia, 3, 50, 56, 155–6, 159
  last natural case of, 172, 218
  in the Middle East, 9, 55
  mortality rate for, 19
  progression of the disease, 7
  in Somalia, 170–1, 218
  transmission of, 8
  Variola major, 12, 69, 151–3, 168, 180, 218
  Variola minor, 9, 12, 69, 151–2, 169, 180
  and variolation, 14–21
  viral stocks of, 4, 173–7
smallpox vaccine, ix–xi, 1, 3, 37–45, 88, 142, 151, 193–4. See also Jenner, Edward; variolation
Society for the Inoculation of the Poor, 43
Solzhenitsyn, Alexander, 167
Somalia, 170–1, 218
Soper, Fred, 154
South Africa, 69, 175
Soviet Union, 3, 127, 135, 154–5, 160, 173, 175–6, 182, 191
Spanish flu, 3, 145–9, 184
Sparham, Legard, 19
Stanley, Wendell, 81
streptomycin, 98
Sumner, James, 91
Supplementary Immunization Activities (SIAs), 204, 212
Sutton method of variolation, 20–1, 30
Sutton, Daniel, 21

Sutton, Robert, 20–1
SV40 virus, 182, 203
swine flu, 148–9, 184
swinepox, 32
syphilis, ix, 32, 41, 60–1, 78, 85, 180

Taubenberger, Jeffery, 148–9
tetanus, 2–3, 61, 82, 85, 91, 106, 114–15, 151, 180–1, 186–7, 193
Theiler, Max, 181
thiomersal, 188–9
Thom, R. A., 33
Thornton, Robert, 43
thrombocytopenia, 183
Thucydides, ix, 8
Tibet, 57
Timoni, Emanuele, 18
Tocktoo, Gilbert, 148
Tronchin, Theodore, 30
trovaflaxin, 202
tuberculosis (TB), ix
  and BCG vaccine, 95–9, 139, 216
  bovine as distinct from human, 95–7
  curative drug treatments for, 2
  discovery of bacteria causing, 85, 92, 94
  as emblematic disease of the 19th century, 91–2
  incidence of, 98
  Jenner's wife's death from, 92, 216
  Jenner's son's death from, 73, 216
  John Keats as sufferer from, 92
  and Lübeck BCG immunization, 95–6
  progression of disease, 92–4, 101
  Robert Koch's work on, x, 85–6, 91–2, 94–7, 144
  and sanatorium movement, 97
  tuberculin skin test for, 91, 94–6
  treatments for, 97–8
  transmission of, 92
tuberculosis (TB) vaccine, 2–4, 95–9, 114, 139, 216
typhoid fever, 16, 31, 81, 85, 91, 96, 114
typhus, 8, 16, 31

Underwood, E. A., 47

United Nations Children's Fund (UNICEF), 193, 198, 204–5, 207, 212–13

vaccination:
arm-to-arm, 38, 49–51, 57, 60–2, 90–1
debate, ix, 65–9, 179–90
defined, x
Jennerian, 48, 57, 59, 62, 65–6, 89–92, 119, 142, 151–2, 156, 177
origin of the term, 39
risk to benefit ration for, ix, 184–6
types of, 89
Vaccination Act of 1840, 58–60
Vaccination Act of 1853, 59–60, 66
Vaccination Act of 1898, 62, 67–8, 90
Vaccine Pock Institute, 35, 44
vaccines:
and adjuvants, 124–6, 129, 141, 150, 187, 194, 196
attenuated, 91, 96, 106, 116, 119, 123–4, 126, 132–7, 151, 177, 183, 189, 194–5, 197
DTP, 186–7
DTaP, 186–7
hepatitis B, 3, 182–3, 185, 193, 197
influenza, 3–4, 96, 117, 119–20, 145, 149–51, 182, 184, 189
MMR, xi, 188
Sabin polio, 133–8, 151, 182–3, 195, 203, 206
Salk polio, 123–38, 183
smallpox, ix, 1, 3, 37–45, 88, 142, 151, 193–4
pertussis, 186–7
TB, 2–4, 95–9, 114, 139, 216
vaccinia, 177, 180, 195
van Leeuwenhoek, Antoni, 79–80
Variola major, 12, 69, 151–3, 168, 180, 218. See also smallpox
Variola minor, 9, 12, 69, 151–2, 169, 180. See also smallpox
variolation:
in Africa, 170
in America, 18–19, 28, 78
banning of, 57–9, 90
of Catherine the Great, 21, 89

in China, 14, 57
dangers of, 19–21, 28, 34, 57–9, 66, 170
defined, 14–15
in England, 14–21, 23, 30, 34, 38–40, 128
in France, 20–1
history of, 1, 14–21
in India, 14, 18
and Jenner's discovery of vaccination, 23, 28, 30, 32–5, 38–44, 48
and Lady Mary Wortley Montagu, 1, 14–15, 50, 170
at Newgate Prison, 16–18, 128
in Russia, 14, 21, 89–90
Sutton method of, 20–1, 30
as test for efficacy of vaccination, 50, 52–3
Vermeer, Jan, 30, 79
Villemin, Jean Antoine, 95
Virchow, Rudolph, 83–4
viruses:
compared with bacteria, ix, 81
defense mechanisms of, 3, 143–4
early understanding of, 32, 42, 77, 81
modern understanding of, 32–3, 81, 83
number of discovered to date, ix
See also specific viruses
Voltaire, François Marie Arouet de, 20–1

Wakefield, Andrew, 188
Walker, John, 51
Wallace, Russell, 66
Walpole, Horace, 112
War of Independence, 28
Warm Springs Foundation, 104–6
Washington, George, 28
Waterhouse, Benjamin, 43, 52–3
Weaver, Harry, 113–14, 117, 120, 123–5, 127
Weller, Thomas, 113
Wellesley, Richard, 56
Wells, H. G., xii
whooping cough, 2–3, 91, 153, 179–80, 186–7, 193
Woodville, William, 40, 44, 48–9
World Bank, 193, 198

World Health Organization (WHO):
  campaign to eradicate polio, 198,
    203, 206–7, 212, 214
  campaign to eradicate smallpox,
    3, 155, 158–60, 165–8, 171–2,
    174–5
  Expanded Program on
    Immunization (EPI), 3, 193, 212
  Millennium Development Goal, 98
World War I, 95, 145
World War II, 3, 87, 111, 145, 151,
  181

Wyeth, Andrew, 138

Yar'Adua, Umara, 203
yaws, 154, 156
yellow fever, 181
Yeltsin, Boris, 173, 176
Yersin, Alexandre, 83
Yersinia pestis, 6. *See also* bubonic
  plague

Zhdanov, Viktor, 154, 174–5